Starting SCIENCE

BOOK TWO

Alan Fraser
Abroath High
Arbroath
Angus

Ian Gilchrist
Kirkcaldy High School
Kirkcaldy
Fife

Oxford University Press

Oxford University Press, Walton Street, Oxford OX2 6DP

Oxford New York
Athens Auckland Bangkok Bombay Calcutta Cape Town
Dar es Salaam Delhi Florence Hong Kong Istanbul Karachi
Kuala Lumpur Madras Madrid Melbourne Mexico City Nairobi
Paris Singapore Taipei Tokyo Toronto

and associated companies in
Berlin Ibadan

Oxford is a trademark of Oxford University Press

© Alan Fraser, Ian Gilchrist 1986

Additional contributions: David Coppock

First published 1986

Reprinted 1986, 1987, 1988, 1989, 1990, 1991, 1992 (twice), 1993,
1994, 1995

Printed with revisions 1996

ISBN 0 19 914241 6 (school edition)
ISBN 0 19 914298 X (bookshop edition)

Printed in Spain

Contents

Contents

Contents

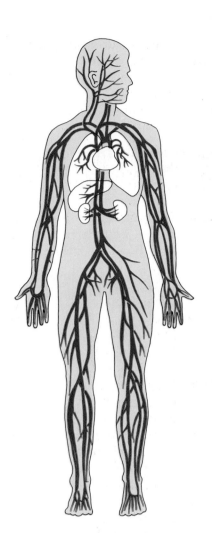

About this book . . .

Starting Science 2 follows on from **Starting Science 1**. It's another scientific book, written to help you
- to understand what you find out in your own experiments
- to see where science fits into everyday life
- to realise how important science is, and how much scientists have been able to improve the world you live in
- to think like a scientist.

In this book, you will meet familiar things – electricity, metals, your body, heat, fuels. But the book also contains much that is unusual. You will read about a train which rides above its rails, a balloon which flew across the Atlantic, an animal which can look in two directions at the same time, and another which can live without drinking. You will learn about red rain, the Space Shuttle, movie cameras, parachuting, building the Pyramids, mountains which were made under the sea, and much more. And, all through the book, you will find questions to make you think and puzzles and cartoons for you to enjoy.

As you can see, **Starting Science 2** is quite a mixture. We hope that the mixture is right, and that you find the book helpful and interesting.

Alan Fraser
Ian Gilchrist

And how to use it . . .

Starting Science is made up of units. Most units contain three pages.

Starting Off is the first page. In it, you will learn a new piece of science. You *must* begin with this page. Otherwise, the other pages won't make any sense.

Going further is the second page. It follows on from what you learned in **Starting Off**.

For the Enthusiast, the third page, takes you even further. The material on it is usually more difficult.

When you start to work on a page, you should first read everything thoroughly – including *Did You Know*. You should also look carefully at any diagrams. Then you can answer the questions. Some questions end with a triangle sign (▲). This tells you that the answer to the question is written somewhere on the page. Some questions begin **Try to find out**. You will usually have to look through other books – like encyclopaedias – for the answers to these. To answer the other questions, you will have to use what you have learned on the page, and a bit of brain power! Using your brain is all part of **Starting Science!**

Heat on the move

Heat can flow from one place to another. You don't have to convince the glassworker in the photograph of that! The molten glass inside the furnace is at a temperature of more than 1000 °C. It gives off a tremendous amount of heat.

This heat could make the glassworker's job very uncomfortable, and very dangerous. In fact, his job *would* be highly uncomfortable and dangerous if he were to:

- wear ordinary clothes
- work with bare hands
- stand directly in front of the furnace, or too close to it.

That's why, when he is working, he is very careful to protect himself.

- He wears special clothes. (His aluminium suit and padded gloves cut down the amount of heat which reaches him.)
- He uses a long metal rod to work the glass. (The heat does flow along the metal, but if he uses a rod made of the correct kind of metal, the heat won't flow along too quickly.)
- He doesn't stand in front of the furnace for longer than necessary.

These things cut down the amount of heat which reaches him, and help to keep him cool. His suit is water-cooled. That helps, too!

There's a lot to learn about how heat flows, and about how to prevent heat from flowing. That's what this section is about.

1

9.1 Heat on the move [1]: conduction

Starting-off

Conductors...

Heat can travel through solids. If you have ever tried to stir boiling soup using a metal spoon, you will know that. Heat energy quickly flows from the soup, through the spoon, to your hand. Your fingers begin to get hot!

The movement of heat through the solid metal is called **conduction**.

The metal spoon **conducts** heat from the hot soup to your hand.

The metal is called a good **conductor** because heat can flow easily through it.

You can stir boiling soup with a metal spoon   but not for long

...insulators...

If you do have to stir hot soup, it's much more sensible to use a wooden spoon, or a spoon with a plastic handle. Like most other non-metals, wood and plastic don't allow heat to flow easily through them. Substances like these are called **insulators**.

There are several insulators (and conductors) in use in this picture. It should help you to answer question 6.

...and their uses

Conductors and insulators can both be useful, but for different jobs. The bottom of a saucepan may be made from a metal, like iron or aluminium, which lets heat flow quickly from the cooker to the food inside. But its handle is likely to be wooden or plastic. These materials won't let heat flow to your fingers and burn them.

1 What is meant by saying that: a) the metal spoon conducts heat b) wood is an insulator? ▲
2 a) Which substances are conductors and which are insulators? ▲
 b) Pick out the conductors and insulators from this list: *aluminium wool glass tin iron cork plastic air*
3 Why is: a) the base of a saucepan made of metal b) the handle made of plastic? ▲
4 How could you protect yourself if you had to lift a hot plate? Why would this protect you?
5 Have you used the words conductor and insulator before? If so, where did you use them? What did they mean?
6 **Try to find out:** where heat insulators are used in your home.

Did you know?

- Copper and silver are the best heat conductors. Copper conducts ten times better than iron. Gases are the poorest conductors. In other words, they are the best insulators.
- When heat travels by conduction the heat is passed on from one atom to the next. The heat flows, the heated atoms don't!

9.1   Trapped air

It's hard to believe, but it's true. The man in the photograph is kept warm by the holes in his string vest! The air in the holes is what matters. When the man pulls on his shirt, it traps air in the holes in the vest. The air is an excellent insulator. It prevents his body heat from escaping and so keeps him warm.

A string vest may seem to be a rather strange piece of clothing, but, in fact, all of your clothes keep you warm in the same way. Clothes have tiny pockets of air trapped between their fibres and so act as insulators. This insulation is important as you have a heat-loss problem. Your body is warmer than the air around you and so it is always losing heat. The insulation supplied by clothes cuts down the amount of heat which escapes and so saves valuable energy.

For a long time, the warmest clothes were made out of natural materials like wool, fur and feathers. These all have tiny pockets of air trapped in them and are excellent insulators. But now man-made fibres are often used instead. A warm anorak is likely to be filled with thousands of fibres made of a plastic like polyester – with air trapped between them.

Here are some other man-made, air-filled insulators.

The holes in the string vest actually help to keep the man warm

This flask can keep drinks warm for 5 hours. It contains an insulating layer of hard plastic foam with air bubbles trapped in it.

Many sleeping bags are filled with man-made fibres. The warmest are filled with a special hollow fibre. It actually has air trapped inside it.

This swimming pool cover is made of plastic sheeting with bubbles of air trapped in it. The air helps the cover to float. It also stops the water from cooling down by loss of heat.

1   Why is a string vest a good insulator? ▲
2   What is the 'heat-loss problem' which affects you (and other mammals and birds)? How do you overcome this problem? ▲
3   Make a list of all the air-filled insulators mentioned on this page. ▲
4   What kinds of 'natural insulation' do birds and mammals have?
5   What makes an air-filled swimming pool cover useful? ▲
6   Suggest why: a) birds fluff out their feathers on a cold day
    b) you should 'fluff up' a sleeping bag before using it
    c) mountaineers who are lost in blizzards dig snow-holes for warmth.
7   **Try to find out:** how Arctic animals survive in the winter.

Did you know?

- All mammals and birds lose heat. But, apart from humans, all have enough 'natural insulation' to allow them to survive.
- Snow is a good insulator. It has air trapped in it.

Aluminium is used for making saucepans, and it's not difficult to see why. Aluminium:

- is a good conductor (and so lets heat flow easily from the cooker to the food)
- has a high melting point (and so does not melt on the cooker)
- is not much affected by chemicals or corrosion.

Aluminium is the 'right material for the job'.

Here are five jobs, and five materials to do them. Match up each material with the job it does. (It will help to decide first of all whether an insulator or a conductor is needed.) Then explain how you made your choice.

The jobs

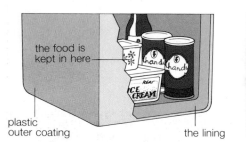

A. Lining a picnic box. The lining must prevent heat from outside the box from getting to the cold food inside.

B. Making a glass blower's blow pipe. The material must be a metal, but it must allow the glass blower to blow glass at over 1000 °C without burning his lips.

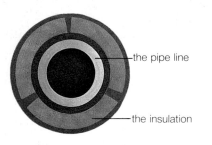

C. Insulating an underwater oil pipeline. The oil is hot when it comes out of the ground. If it cools too much, it won't flow.

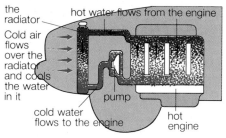

D. Making a car radiator. The radiator has to cool hot water from the car's engine. The water is passed through the radiator which has air flowing over it.

E. Making a spacecraft's heat shield. The shield has to protect the spacecraft's metal body from burning. Fast moving spacecraft get very hot when entering the Earth's atmosphere.

The materials

(Key: ● = conductor
         ○ = insulator
The more dots the better)

Iron ● ● ●
Metal. Melts at 1535 °C
Strong, but rusts. Quite cheap

Ceramic fibre ○ ○ ○ ○ ○
Fibres made of special glass-like material. Can be made into blocks and tiles. Very high melting point. Fire proof. Extremely expensive

Expanded polystyrene ○ ○ ○ ○
Plastic foam. Keeps its shape but is not strong. It can be squashed easily. Melts and burns easily. Cheap

Copper ● ● ● ● ●
Metal. Melts at 1083 °C
Quite strong. Does not rust
More expensive than iron

PVC foam ○ ○ ○ ○
A plastic foam. Very hard wearing. Waterproof. Not easily squashed. Burns and melts easily

9.2 Heat on the move [2]: convection

Heat energy can travel through liquids. This *must* happen when you boil water in an electric kettle. The kettle's heating element only warms the water next to it, but the heat energy is carried all through the water until it boils.

Heat energy can travel through gases. This *must* happen when you heat a room using an electric convector heater. The heater only warms the air inside it, but soon you can feel the warmth all through the room.

The heat can't be travelling through the water or air by *conduction*. (Water and air are both bad conductors.) The heat travels in a different way, by **convection**.

How heat energy travels through the water in the kettle

→ heated water moves in this direction
→ cold water moves in this direction

How convection works.

When the convector heater is switched on, it sets up a flow of air in the room. This flow is called a **convection current**.

The convection current is set up by cold air sinking through warm air. (Cold air sinks because it is more dense than warm air.)
When the heater is switched on:
1 the heating wires heat the air round them
2 the hot air rises and escapes from the top of the heater
3 cold air is drawn in to the bottom of the heater to take its place.

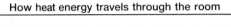

How heat energy travels through the room

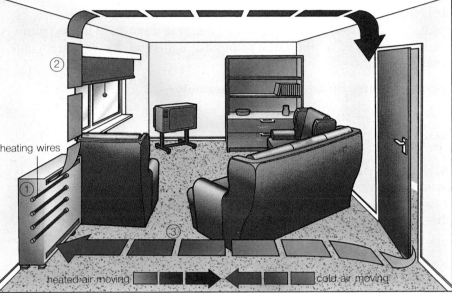

heating wires

heated air moving   cold air moving

Switching on an electric kettle also sets up convection currents, in the water inside. (Hot water rises from the heating element and cold water sinks downwards to take its place.) In fact, convection is the main way in which heat travels through all liquids and gases. It takes place whenever one part of a liquid or gas is heated more than the rest. The colder liquid (or gas) sinks. The heated liquid (or gas) rises to take its place.

How air moves round about a fire

→ heated air moving
→ cold air moving

1   Copy and complete:
   a) Heat travels through _____ and _____ by convection.
   b) Convection takes place whenever ...
   c) Heat can't travel through solids by convection because ... ▲
2   a) Explain how a convector heater warms all the air in a room. ▲
   b) If you fix paper decorations to the wall above a convector heater, they flutter when the heater is switched on. Why does this happen?
3   Why is it difficult to heat rooms with high ceilings?
4   On which shelf of an oven do cakes cook quickest, and why?
5   The wall above a radiator is often dirty with dust from the floor. How does the dust get there?
6   **Try to find out:** why firemen enter smoke filled rooms by crawling.

Did you know?

- Smoke rises by convection. When the air above a fire is heated, it rises, carrying tiny particles of ash with it.
- Convection can't take place in solids. It only takes place where the atoms and molecules can move once they are heated.

9.2 Flying on hot air

Have you ever wondered what keeps a glider in the air? Gliding does seem to be a bit of a mystery. A glider has no engine to power it, yet it can fly for long distances before landing. It can even climb through the air.

The glider stays in the air, and climbs, because of air currents. At the beginning of each flight, the glider has to be towed into the air. Once it has been released, however, it can fly for as long as the pilot can find rising air currents to keep it airborne. The best pilots are highly skilled at finding these air currents. Cross-country flights of 300 km or more are often carried out.

Below you can see a diagram of a glider's cross-country flight. You can see where the glider climbs, and where it loses height. Most of the rising air is in the form of **thermals**, currents of hot air rising because of convection. A thermal is produced whenever some air is heated more than the air round about it. If a pilot finds a thermal, he will circle in it and gain height.

A cross-country flight – thanks to thermals

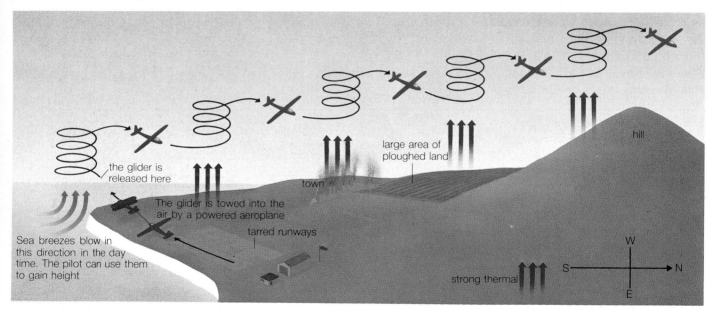

the glider is released here

The glider is towed into the air by a powered aeroplane

tarred runways

Sea breezes blow in this direction in the day time. The pilot can use them to gain height

town

large area of ploughed land

hill

strong thermal

1 a) What is a thermal?   b) When is a thermal produced? ▲
2 How does a glider: a) get off the ground   b) gain height when it is flying? ▲

About the cross-country flight in the diagram

3 Where does the glider rise, and why?
4 Suggest why thermals rise from the town.
5 The Sun heats the ground. The ground heats the air above it. Using information on the diagram, work out whether dark or light ground takes in heat better. Give reasons for your answer.
6 From which side of the hill (North or South) do thermals rise? Why do they rise from this side?
7 Imagine that you were the glider pilot. Write a story to describe your flight – what happened, and what you saw.

Did you know?

- Birds make use of thermals to soar in the air. South American condors can fly to heights of 4 km using thermals.
- Glider pilots often find thermals by watching birds in flight.

6

9.2 Convection in nature

The Monsoon

What can make a wind blow? Convection can. Whenever hot air rises from the Earth, cold air flows over the Earth's surface to take its place. This moving air is a wind.

The **Monsoon** is a wind which affects much of Asia. In summer, it is a warm, rain-carrying wind which blows from the sea to the land. In winter, it is a dry cool wind, blowing from land to sea.

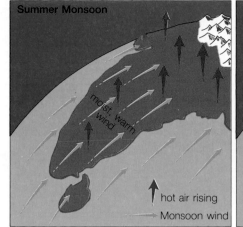
Summer Monsoon

moist, warm wind

hot air rising

Monsoon wind

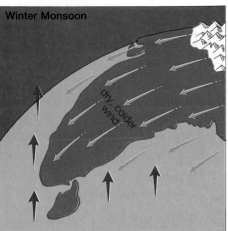
Winter Monsoon

dry, colder wind

The Monsoon is a 'convection current wind'. It blows because of temperature differences between the land and the sea. In summer, the land heats up more quickly than the sea. Cold air flows in from the sea and warm air rises above the land. In winter, the land cools down more quickly than the sea. Cold air from the land flows out to sea pushing warm air upwards above the sea.

Red rain

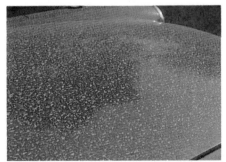

Occasionally, coloured rain falls in Europe. Convection currents in North Africa are to blame for this. Strong currents of rising hot air carry desert dust high into the atmosphere. The dust is carried Northwards by the wind, then falls with the rain.

'Upside-down convection'

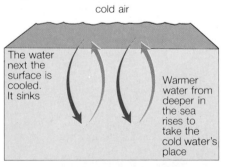

cold air

The water next the surface is cooled. It sinks

Warmer water from deeper in the sea rises to take the cold water's place

'Upside-down convection' takes place in the sea in winter. Then the sea is warmer than the air above it. The water at the surface loses heat to the air and cools down. The cooler, denser water sinks to the bottom of the sea and warmer water rises to the surface.

Cooling by volcano

Erupting volcanoes produce huge convection currents which carry volcanic ash and dust high into the atmosphere. As this dust spreads out, it forms a 'blanket' over parts of the Earth. This cuts down the amount of sunshine which reaches the Earth's surface, and lowers temperatures.

1 What is a wind? ▲
2 a) Where does the Monsoon blow, and what causes it? ▲
   b) Why does the Monsoon blow northwards in summer and southwards in winter? ▲
3 Explain how a) upside-down convection affects sea temperatures in winter b) volcanoes affect the weather. ▲
4 Using the photograph to help you, draw a diagram to show how convection currents are produced at a volcano.
5 **Try to find out:** where Trade Winds blow, what causes them and which sailing ships used them 130 years ago.

Did you know?

● The summer Monsoon usually brings very heavy rain. In 1915, the Indian town of Cherrapungi had 90 cm of rain in 24 hours!

9.3 Heat on the move [3]: radiation

When you toast bread under a grill, some heat travels **downwards**, from the grill to the toast.

This heat can't be carried by *convection*. Convection carries hot air upwards.

The heat can't have been *conducted* to the bread either. The air between the bread and the grill is a poor conductor.

The heat is travelling by **radiation**.

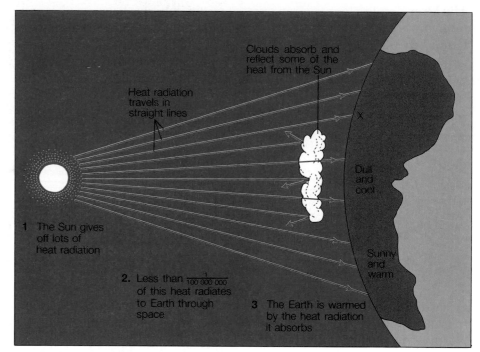

The grills heating element gives off heat in all directions

Some of the heat travels **downwards** to the bread and toasts it

Clouds absorb and reflect some of the heat from the Sun

Heat radiation travels in straight lines

X

Dull and cool

Sunny and warm

1 The Sun gives off lots of heat radiation

2. Less than $\frac{1}{100\,000\,000}$ of this heat radiates to Earth through space

3 The Earth is warmed by the heat radiation it absorbs

Earth-warming radiation

The Earth is warmed by heat radiation. This comes from the Sun. (The Sun also gives off other kinds of radiation, including light.)

Only a small fraction of the Sun's heat radiation reaches the Earth. To get there, it travels through space. Since space has very few particles in it, this shows that **heat radiation is not carried by moving particles**.

When the heat radiation reaches the Earth, some of it is taken in, or **absorbed**. Some is bounced back, or **reflected**. It's the absorbed heat which warms up the Earth. **The more heat absorbed, the hotter it gets.**

Clouds can cut down the amount of heat radiation reaching the Earth. Cloudy days are usually cool days! That's because **heat radiation travels in straight lines**. The heat can't bend round the edges of the cloud to reach the Earth underneath.

1 Explain why heat radiation can't be travelling from the grill to the toast by conduction or convection. ▲
2 What is meant by: a) absorbing heat b) reflecting heat? ▲
3 What happens to an object when it absorbs heat? ▲
4 Why can heat travel through space by radiation, but not by conduction or convection?
5 Give evidence which shows that heat travels in straight lines.
6 What's the weather like at X (on the diagram)?
7 a) What is infra-red radiation? ▲
   b) **Try to find out:** some uses of infra-red detectors.

Did you know?

- Heat radiation is also called **infra-red radiation**.
- Everything gives off heat radiation. The hotter the object, the greater the energy of the radiation.
- Burglar alarms can be set off by heat radiating from a burglar's body.

9.3 Heat out, and in

Giving heat out . . .

This is a heat picture called a **thermogram**. It was taken using a special camera which detects heat rays.

The thermogram shows someone drinking out of a mug. (You can see a normal photograph of the person above.) The colours on the thermogram show how much heat is being given out. Use the colour code, and the thermogram, to answer the questions.

1 Which part of the person gives out: a) most heat b) least heat?
2 Why do some parts of the person lose more heat than others?
3 Which gives out more heat, the person's head or arms?
4 Is it a hot or cold drink? Explain your answer.

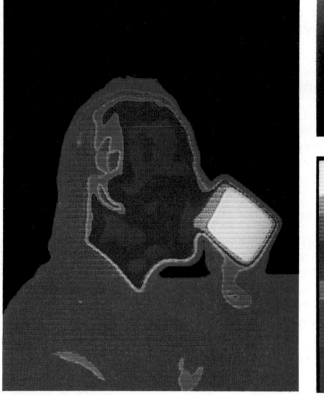

above 38·6 °C

35·3 °C

32·3 °C

29·4 °C

26.5 °C

less than 23·5 °C

. . . and taking heat in

Some surfaces absorb heat better than others. The diagrams on the right show that. They show what happens to heat from the Sun when it strikes different surfaces.

Key:

 heat from Sun

heat reflected from surface

heat absorbed by surface

40%, 60% etc   percentage of heat absorbed or reflected

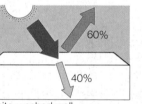
white-washed wall — 60% / 40%

dull aluminium — 50% / 50%

earth — 15% / 85%

red-brick wall — 30% / 70%

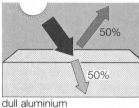
polished aluminium — 75% / 25%

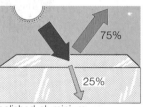
tar — 10% / 90%

1 Which surfaces absorb heat best: a) dull or shiny b) dark or light? Explain your answers. ▲
2 Copy and complete the bar chart to show how well the different surfaces reflect heat. Which reflect heat best?
3 Why is shiny metal fixed to the back of an electric bar fire?
4 You can melt snow quickly by putting soot on it – if there is some sunshine. Why does this work?
5 In very hot countries, houses have white-washed walls. Why is this useful?
6 **Try to find out:** why skiers use lots of sun tan cream.

Percentage of heat reflected — 100%, 80%, 60%, 40%, 20% — white-washed wall, red-brick wall, earth

9

9.3 Two tone balloons

How can the balloon be kept in the air? That's the problem facing every balloonist who wants to make a really long flight. Giving the balloon a two-tone, silver and black colour scheme doesn't solve the problem, but it certainly can help!

First across the Atlantic – Double Eagle II

The first balloon to cross the Atlantic (Double Eagle II) had a silver and black colour scheme. It was crewed by three Americans, Ben Abruzzo, Max Anderson, and Larry Newman. The crew, their equipment and bags of sand ballast were carried in the balloon's **gondola**, a small steel and fibreglass boat. The gondola and its contents were kept in the air by the balloon's **envelope**. This was made of nylon, and filled with helium, a gas much less dense than air. The helium gave the balloon its lift.

On August 11th, 1978, with weather forecasters predicting air currents moving steadily eastwards over the Atlantic, the balloon took off from Maine, U.S.A. Getting the balloon into the air was relatively easy. Keeping it flying evenly, however, was much more difficult. Although the size of the envelope had been carefully matched to the gondola's weight, the balloon's height changed frequently. Temperature changes were to blame. The heat of the day made the helium expand and become less dense. This increased the balloon's lift, causing it to rise. At night, the gas cooled, contracted and became more dense. The balloon fell.

In order to keep the balloon in the east-moving air, its height had to be controlled. When the balloon rose too far, helium was allowed to escape through a valve. When it fell too far, sand ballast was thrown overboard to lighten the gondola. Unfortunately, each cubic metre of helium released and each sand bag thrown overboard brought the end of the flight nearer. That's why the colour scheme was important. The silver upper part of the envelope reflected the Sun's heat during the day, preventing the helium from overheating. The black lower part of the envelope absorbed heat radiating from the Earth at night, and from the early morning Sun, preventing it from cooling too much. This saved precious helium and ballast, prolonging the flight.

Helped by the colour scheme, Double Eagle II did stay in the air long enough to make the crossing. 137 h, 5 min, 50 s after take off, 5001 km from its launch, it landed in France.

1 What was each of the following used for in Double Eagle II?
   helium   the gondola   sand ballast   envelope ▲
2 What kept the balloon moving across the Atlantic? ▲
3 Why did the balloon's height change: a) in the day
   b) at night? ▲
4 Why did the balloon's height have to be controlled and how was it controlled? ▲
5 Why was the balloon's colour scheme important? ▲
6 Suggest why: a) they used a boat as a gondola   b) the balloon almost crashed into the sea after entering a cold cloud.

Did you know?

- The crew had to jettison their ballast and much more to stay in the air. They threw overboard radios, a life-raft, parachutes and $30 000 worth of equipment.
- Anderson and Abruzzo had made an earlier unsuccessful attempt in 1977. They were rescued from the sea off Iceland.

9.4 Don't let the heat escape!

The Blacks, the Browns, the Greens, and the Whites had two things in common. They lived in the same type of house, and had huge heating bills.

One year, they each had heating bills of nearly £800. That made them think!
They decided that they would have to do something to reduce these bills. And so they decided to insulate their homes to cut down the amount of heat energy which was escaping.

Each family used a different kind of insulation.

Insulating walls with foam

The Blacks put insulation in the loft. They laid shiny aluminium foil above the ceiling, then covered it with glass fibre wool 7.5 cm thick. It cost them £200 to do this. Their heating bills went down by £100 each year.

The Greens fitted double glazed windows everywhere. These windows have two panes of glass with a layer of air trapped between them. This cost them £2500. Their bills went down by £100 each year. The house was quieter, too.

through the roof
25%

through the windows
10%

through the wall
35%

through the floor
15%

in draughts
15%

Where the heat escapes

The Browns spent £100 cutting down draughts. They fitted draught excluders on all doors and windows, and blocked up fireplaces which were never used. This saved them £50 each year.

The Whites insulated the walls. Each wall was made of two layers of brick with an air space in between. The Whites filled this space with insulating foam at a cost of £700. Their bill went down by £200 each year.

1 List: a) 4 ways in which heat can escape from a house
   b) 4 ways of preventing heat from escaping. ▲
2 Which methods of insulation use trapped air as an insulator?
3 Why is aluminium foil fixed above ceilings?
4 Explain how heat can escape through an unused fireplace.
5 How much did the Blacks: a) pay for insulation b) save each year? ▲
6 a) The Blacks only really began to save money after two years. Explain this.
   b) How many years passed before the other families began to save?
7 If you had to insulate a house, which two methods would you use first, and why?

Did you know?

● One experimental house in Wales is so well insulated that its 17 rooms can be heated using only a 3 kilowatt heater. (This heater produces heat at the same rate as a 3-bar electric fire.)

9.4 Saving body heat

The most important heat to save is your body heat. If your body gets too cold, it won't work properly. Then you are said to be suffering from **hypothermia**.

You are unlikely to suffer from hypothermia — as long as you have shelter, enough clothes to wear (and, of course, enough to eat). But some people, like divers, mountaineers and astronauts, live and work in more extreme conditions. For them, hypothermia is a real problem. They have to wear special clothes to survive.

For divers, the problem is caused by the cold sea water. The moving water quickly carries heat away from the skin.

Those divers who work in shallow waters wear **wet suits** to stay warm. These suits are made of rubber which has bubbles of air trapped in it. Surprisingly, the suits also use water as an insulator. A thin layer of water is trapped between the suit and the diver's skin. Being a poor conductor, this water helps to prevent the diver's body heat from escaping.

Mountaineers face dangers from sub-zero temperatures, and from driving rain and snow. If their clothes become soaked through, they quickly become chilled. They normally wear waterproof outer clothing and heavily insulated inner clothing to help them survive. Until recently, however, this arrangement of clothing did not allow water vapour (from sweat) to escape, and so the mountaineers' clothes became wet with condensation.

Modern fibres solve this problem. The most up-to-date waterproof clothing allows water vapour to escape. The newest insulating fibres soak up very little water.

Astronauts face all sorts of dangers. A 'space suit' contains several layers, each of which protects the astronaut from a particular danger. There are layers for protection against fire, cosmic rays, and flying particles. There are also several insulating layers which protect the astronaut against the extreme temperature which he meets outside the spacecraft. In fact, the astronaut is so well insulated that his body heat can't escape. He needs a cooling system to prevent overheating! He carries a small refrigerator on his back. Cold water from this refrigerator flows through small pipes in the astronaut's underwear and takes away some of his body heat.

Skin diver

Mountaineer

Astronaut

1 What is hypothermia? How can it be prevented? ▲
2 Why can eating food help to prevent hypothermia?
3 Why is hypothermia a real problem for: a) divers
   b) mountaineers? ▲
4 Write down some of the dangers facing an astronaut on the Moon. ▲
5 Why does the astronaut need a cooling system? How does it work? ▲
6 If you fall into the sea, the water quickly cools your body. If you swim in a wet suit, water keeps you warm. Explain this difference.
7 **Try to find out:** a) how long distance swimmers keep warm
   b) why women survive low temperatures better than men.

Did you know?

● During the day, the temperature on the Moon can be 120 °C. At night, it can be −155 °C.
● The astronaut's body has to be kept under pressure in his space suit. Without this pressure, his blood would boil!

9.4 Making the most of a free gift

In a time of energy shortage, you aren't likely to be impressed by a home which loses 99.9% of the heat supplied to it. And yet your home – the Earth – does just that. Every day, a huge quantity of heat energy reaches the Earth from the Sun. Much of this heat is absorbed, warming up the land and the sea, but then, at night, practically all of the absorbed heat is lost again. It escapes into space as radiation while the Earth cools down.

With so much free energy being wasted, it's not surprising that scientists are trying to find ways of using and storing large quantities of **solar heat**. A number of methods of trapping this heat have been developed, but most of them suffer from one main disadvantage. The equipment used is expensive, which makes the large scale use of solar heat a costly business.

Despite this, solar energy is used for heating homes and other larger buildings in many parts of the world. Even in Britain, where sunshine can hardly be guaranteed, solar heating is being used in many homes. The homes which use solar heating best are those which have been specially designed for that use. They have been built:

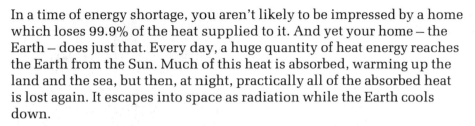

A solar cooker. It reflects the Sun's rays on to the pan

to face north and south so that one side of the house receives sunshine most of the day.

with large windows facing south. The Sun's heat rays travel through the glass and are absorbed inside the home.

with a large greenhouse on the south side. The air in the greenhouse heats up. It can then be circulated through the house.

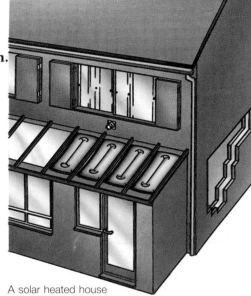

A solar heated house

with solar panels on the roof. Water from the house hot water system trickles through the panels. As it does so, the Sun heats it up.

with shutters on the windows to prevent heat from escaping when the air outside cools.

with very thick walls which warm up during the day and radiate some of the heat into the house at night.

and, of course, with masses of loft insulation.

1 Why does the Earth's temperature:
   a) rise during the day and fall at night ▲
   b) change more on clear days (and nights) than on cloudy ones?
2 a) Why is so little use made of solar energy? ▲
   b) Suggest two reasons why Japan, U.S.A., and Israel have more solar heated homes than any other country.
3 Why does the solar heated house: a) face North and South
   b) have thick walls   c) have shuttered windows   d) have solar panels? ▲
4 a) What colour should solar panels be, and why?
   b) In what direction should solar panels point, and why?
5 **Try to find out:** one other way of using solar energy.

Did you know?

- The solar energy reaching the Earth in 1 year is more than 10 000 times greater than the world's energy needs.
- Solar panels can absorb useful amounts of heat even on cloudy days.

13

Hot but no hotter

It's very likely that the classroom where you are working just now is heated by part of a very large central heating system. Somewhere in the school, there will be a big boiler where water is heated by burners which use oil or gas. The hot water is piped to radiators in the classroom. There it gives out the heat to keep you warm.

Like every workplace, a school has to be run at a comfortable temperature. In fact, the law sets a minimum classroom temperature. Architects have to do some careful calculations when they are designing a school. They have to make sure that the boiler can supply enough heat to keep all the classrooms at the correct temperature even in the coldest weather. Heat energy is measured in **kilojoules:**

● **1 kilojoule is the amount of heat energy needed to heat up 1 kilogram of water by 1 °C.**

A small school will need a smaller boiler than a large school.

You may wonder why different sizes of boiler have to be used. After all, all schools have to be heated to the same temperature. The answer is that, in a large school, there are more rooms, more radiators and so more water to heat. If the boiler isn't big enough it won't be able to heat the water quickly enough and the rooms will be cold.

But as you well know, the room can get too hot. If the radiator keeps pumping out heat energy, the room temperature will keep going up, and so will yours. That wastes heat (and money) and makes the room uncomfortable. That's why **thermostats** are used. Their job is to keep the temperature of the room steady. They contain sensors which detect whether the room is hotter or colder than it should be. When the temperature is too high, thermostats shut off the flow of hot water, turning it on again when the temperature gets too low.

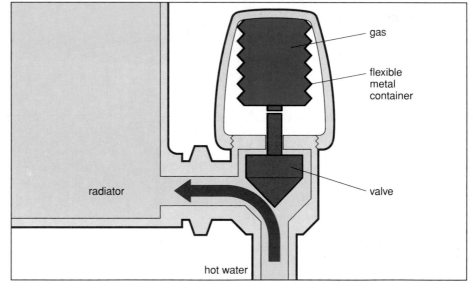

When the gas becomes hot the container expands and pushes the valve down and turns off the flow of hot water. When the gas cools the container contracts and the valve rises to let hot water flow again.

1 What is a thermostat's job? How does it work?
2 Which units are used to measure: a) heat b) temperature?
3 Which has: a) higher temperature b) greater amount of heat; a bucket of water at 100 °C, or a hot needed at 500 °C? Explain.
4 What can be done to prevent:
a) heat being lost when water flows through pipes
b) a room from getting too hot?
5 **Try to find out:** how your school heating system works. Then try to find ways of saving heat in the school.

Did you know?

● The highest temperature for the coals of a 'fire-walk' has been recorded at over 800 °C. (**Don't try it!**)

Hydrogen, metals, acids and alkalis

Hydrogen gas, a **metal** and an **acid** all had a part to play in the successful flight of this pioneer balloon. The hydrogen was used to fill the balloon skin and so keep the balloon in the air. The metal and the acid were both used to make the hydrogen. It was made in a barrel by mixing iron with sulphuric acid and was fed from there along a pipe into the balloon.

The balloon was one of the earliest flying machines. It was designed and built by a Frenchman, Jacques Charles. On 27th August, 1783, one took off from a park in the middle of Paris. Thousands of people watched the take off, but they didn't see very much. The balloon rose in a shower of rain and soon disappeared from view. Three quarters of an hour later, it landed 10 kilometres away. The only people to see it land were some terrified farmers. They thought that it was a monster, and attacked it with pitchforks and scythes!

Hydrogen gas, **metals**, and **alkalis** have important jobs to do in this flying machine – the Space Shuttle. The hydrogen is used to get the Shuttle off the ground. It's the main fuel burned by the engines. The metals are used in the Shuttle's body. (Its skin, for example, is mainly aluminium.) The alkalis are used to remove carbon dioxide from the air breathed out by the astronauts, and also in the Shuttle's batteries.

The Shuttle is one of the most modern space vehicles. It was built by American scientists and engineers. Hundreds of thousands of people watched it take off on its first mission from Cape Kennedy, U.S.A. on 12th April, 1981. Millions more watched on T.V. Within 2 minutes the Shuttle had blasted its way to a height of 40 kilometres. 54½ hours later, after orbiting the Earth 36 times, and travelling 1 729 318 kilometres, it glided back to Earth. It landed in front of another huge crowd – to congratulations from everyone.

Hydrogen, **metals**, **acids**, and **alkalis** are important chemicals, and not just for flying machines! In this section you will learn more about them, and some other reasons why they are so important.

In this photograph, you can see a weather balloon. It's being filled with gas. Once it is filled, it will be released.

The balloon's job is to carry instruments high into the atmosphere. The instruments make measurements of the weather. Then they send the information back to Earth.

Not every gas is useful for filling weather balloons. The gas must be much less dense than air, otherwise the balloon won't lift the heavy instruments off the ground.

The gas normally used is **hydrogen**. It's ideal for this. It's much less dense than air, and quickly floats upwards when released. In fact, hydrogen has the smallest density of all substances.

A test for hydrogen

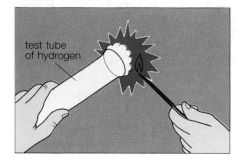

A test tube of hydrogen seems to be empty. Hydrogen gas doesn't have a colour. (It doesn't have a smell either!) But if you put a burning splint to the mouth of the test tube, you will know that there is something in it. **Hydrogen burns with a pop when mixed with air.** This test allows you to pick out hydrogen from oxygen, nitrogen, and carbon dioxide.

Water from hydrogen

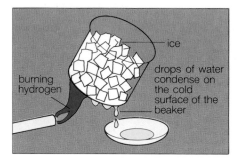

If you put a jet of burning hydrogen onto a cold surface, a colourless liquid collects. The liquid boils at 100 °C. In other words, the liquid is water.

When hydrogen burns, water is made. Water is a compound, made by joining together the elements hydrogen and oxygen. Its chemical name is hydrogen oxide.

Hydrogen from water

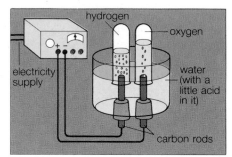

You can use this apparatus to pass an electric current through water. When the current is switched on, bubbles of gas appear on the two rods. (If a little acid is added, the gases are made more quickly. The acid helps the current to flow.)

Hydrogen is made at one rod, and oxygen at the other. The electrical energy splits the water into hydrogen and oxygen.

1 What is hydrogen like? Describe it. ▲
2 a) What is a weather balloon's job? ▲
   b) Why is hydrogen a good gas for filling weather balloons? ▲
3 How can you pick out hydrogen from other common gases? ▲
4 How can you make: a) water from hydrogen   b) hydrogen from water? ▲
5 a) Water is a compound. What does this mean?
   b) What is water's chemical name? Why is it called this? ▲
6 Hydrogen gas in nature is only found high in the upper atmosphere. Why should it be found there?
7 **Try to find out:** why modern airships are filled with helium, not hydrogen.

Did you know?

- Hydrogen means 'water producer'.
- Pure hydrogen makes up less than 0.000 001% of the atmosphere.
- Hydrogen is found in many important compounds. Most of your body is made up of compounds containing hydrogen.

10.1 A useful fuel...

Mixtures of hydrogen and air can explode when they are lit. But if hydrogen is mixed with the correct amount of air, it burns with a hot, clean flame. Hydrogen is a very useful fuel.

Homes and factories can be heated with the help of hydrogen. For many years, British homes and factories were supplied with town gas, containing 50% hydrogen. It was made at the local gas works by heating coke (carbon) with steam. Although North Sea gas is used now, it will run out one day. Then hydrogen could well be used again.

Some welding torches burn hydrogen. When hydrogen is mixed with pure oxygen, it burns with a very hot flame. Hydrogen-oxygen welding torches can produce flames as hot as 4000°C, hot enough to join pieces of metal together.

Hydrogen can be useful as a fuel for transport. But, at present, it is not widely used for this. There are some hydrogen powered buses and cars, but they are mostly experimental. Only space rockets have made much use of hydrogen as a fuel. The American Space Shuttle is a hydrogen powered space craft. Its Orbiter sits on a fuel tank which has two compartments in it. One contains liquid hydrogen which the engines burn. The other contains the liquid oxygen which is needed to keep the hydrogen burning.

A hydrogen-oxygen welding torch gives a flame hot enough for underwater welding

The Space Shuttle and how it works

The Orbiter is designed to go into orbit, carry out its mission and return to Earth. It carries a little fuel.

engines

The fuel tank carries fuel for the Orbiter's engine. The fuel is used up just as the Orbiter enters orbit at a height of 200 km. Then the tank is ejected. It enters the atmosphere and burns up.

Booster rockets help the Shuttle to get off the ground. When the Shuttle has reached a height of 50 km, the fuel in these rockets is completely used up. The rockets are ejected. They parachute back to Earth to be used again.

1 Why is hydrogen a good fuel? ▲
2 Give three uses of hydrogen as a fuel. ▲
3 Why does the Space Shuttle carry: a) hydrogen b) oxygen? ▲
4 Why are clouds of steam produced when the Shuttle takes off?
5 Write down two jobs carried out by the Shuttle's fuel cell. ▲
6 Write down what happens to the Shuttle's booster rockets, fuel tank, and Orbiter on a flight. Then draw a diagram to show what happens on a Shuttle flight between lift-off and orbit.
7 **Try to find out:** about experiments carried out by the Shuttle.

Did you know?

● The Shuttle's electricity is made by reacting hydrogen and oxygen in a special *fuel cell*. At the same time, this cell produces all the water needed by the astronauts.

10.1 ...the fuel of the future?

Hydrogen is the fuel of the future.
That's the claim made by a number of scientists. They think that hydrogen is such a good fuel that it will one day replace all of the fuels used at present. Could the scientists be correct. Judge for yourself as you read on.

A good fuel should produce large amounts of heat when it burns.
Heat energy is measured in **kilojoules (kJ)**. Burning 1 g of hydrogen produces 143 kJ of heat, enough to boil a large cupful of water. (Burning 1 g of petrol produces 48 kJ. Burning 1 g of natural gas produces 58 kJ.)

A good fuel should be cheap and easy to produce.
There is a very small amount of hydrogen gas in the atmosphere, but it cannot be easily separated from other gases.

Hydrogen gas can be produced from water, and there is plenty of that. At present, most hydrogen is made by heating steam and natural gas together, or by passing an electric current through water. Unfortunately, these methods are expensive. More energy is needed to make the hydrogen than could be produced by burning it! A recently discovered method, however, could lead to hydrogen being produced more cheaply. It uses light energy from the Sun to split up the water.

A good fuel should be easily transported and stored.
Hydrogen can be stored as a liquid under pressure in strong tanks. It can be transported from one place to another by pipes. That makes hydrogen useful in industry and for heating homes. But liquid hydrogen storage tanks are really too heavy to carry on vehicles. As a result, the designers of hydrogen-powered buses and cars have had to find other ways of storing the fuel. Most have fitted the vehicles with tanks of **metal hydrides**, compounds containing hydrogen joined to the metal. These compounds give up hydrogen on heating. (Heat from the vehicle's engine can be used to do this.) But, again the compounds are expensive.

Hydrogen-powered vehicles—tomorrow's transport?

A good fuel should be safe to use.
Like all liquid or gaseous fuels, hydrogen can be dangerous. It can leak through tiny cracks in pipes and form an explosive mixture with air. But if hydrogen does leak, it can quickly escape upwards through the air. Hydrogen's **ignition temperature** (the temperature at which it catches fire) is actually higher than for many other fuels. And so hydrogen is safe enough to use.

1 How is hydrogen made? Why is it expensive to make? ▲
2 What are metal hydrides? Why are they useful? ▲
3 Explain why hydrogen is more useful at present as a fuel for heating houses than as a fuel for cars.
4 Does hydrogen cause pollution when it burns? Explain.
5 143 kJ of heat will boil a large cupful of water. How many grams of:
   a) hydrogen b) petrol c) natural gas will supply this heat on burning?
6 How do petrol and hydrogen compare as fuels? What are their advantages and disadvantages?

Did you know?

● A village called 'Hydrogen Homestead' in Utah, U.S.A. depends completely on hydrogen for heating homes and powering cars and tractors.

10.2 How reactive are metals?

When you put a piece of magnesium into water, bubbles of hydrogen gas are produced. The bubbles appear very, very slowly. It could take about a week to collect enough hydrogen to 'pop'.

When you put a piece of magnesium into dilute hydrochloric acid, hydrogen gas bubbles off quickly.

This shows that:

magnesium reacts slowly with water and quickly with acid.

Magnesium in water   Magnesium in acid

A few metals react with water. More metals react with acid. The pictures show what happens when different metals are put in acid and in water. Some metals react faster than magnesium. Some metals react more slowly. Some metals (the **unreactive metals**) don't react at all.

The **reactivity series** is a kind of league table of metals. It puts the metals in order, with the fastest reacting ones first. Potassium is at the top of the league. It reacts very quickly with water. If it is put in acid, there is an explosion!

Silver and gold are at the bottom of the league. They don't react with acid or water. The rest of the metals are in between. You can work out the order for yourself as you answer the questions.

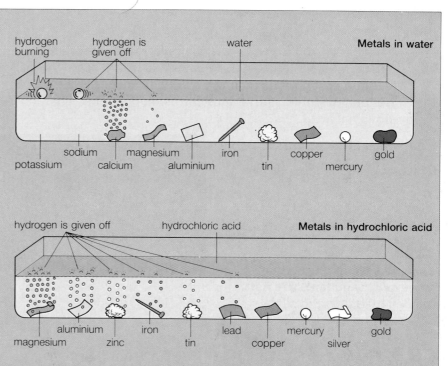

Did you know?

- When potassium reacts with water, so much heat is produced that the hydrogen catches fire.
- Platinum is even less reactive than gold.

1 How could you show that magnesium reacts faster with acid than with water? What would you see? ▲
2 What is: a) the reactivity series b) an unreactive metal? ▲
3 Why should you never put potassium in acid? ▲

Use the information on the pictures to answer questions 4–6

4 Write down: a) two metals which react with water b) two unreactive metals c) two metals which are less dense than water.
5 Would it matter if you a) wore a gold ring while washing dishes b) spilt battery acid on a car body (which is made of iron) c) used calcium for making water pipes? Explain your answers.
6 Copy the table on the right, then use it to make your own reactivity series. (The faster a metal reacts, the higher it should be.)

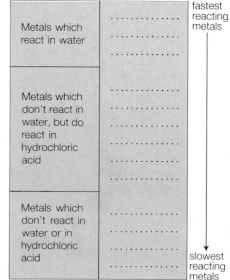

Make your own reactivity series
(Use the diagrams above to help)

10.2 Using unreactive metals

The two nails in the photograph look very different. The top one has been taken straight from the packet. It is still shiny. The bottom one has been in acid for a week. The acid has reacted with the iron in the nail and has started to eat it away. The nail has **rusted** or **corroded**.

A metal corrodes whenever a chemical attacks its surface. Water, air, and acid are three of the most common corroding chemicals. Acid fumes from burning coal and oil dissolve in rain water to make **acid rain**. Acid rain causes lots of damage by corroding metals. Metals first lose their shiny surface, then, as more is removed, they lose their strength.

Metal corrosion is a big problem. It is most serious for the metals at the top of the reactivity series. They corrode so rapidly that they are of little use. It is also serious for less reactive metals like iron. The rusting of iron costs millions of pounds each year. Only the metals at the bottom of the series do not suffer from serious corrosion problems. They corrode slowly, if at all.

It's easy to tell which nail has been in acid

Using unreactive metals

Metals which don't corrode can be very useful.

Gold is used for making jewellery. It does not corrode and so stays shiny.

Copper is used in plumbing. Copper pipes and tanks are not corroded by water, and so they don't leak.

Tin is used for coating tin cans. The can is really made out of steel (which is mostly iron). This makes the can strong and cheap. The tin protects the steel from being corroded by the food in the can.

Titanium is a metal which has recently become important. It is strong and light and is not easily corroded. Most titanium is used to make aircraft. It is also used for making artificial hip joints.

1 a) What happens to a piece of metal when it corrodes? ▲
   b) Write down three things which can make a metal corrode. ▲
2 a) Where in the activity series are the metals most affected by corrosion? ▲
   b) Name some of these metals, and some which don't corrode.
3 Why does a 'tin' can have tin and steel in it? ▲
4 a) Name some metals which are used to make jewellery.
   b) Why are these metals used?
5 Why is it dangerous to drive a car which is really rusty?
6 **Try to find out:**   a) why metals corrode more quickly near the sea
   b) why car exhausts rust so quickly.

Did you know?

- The coins in the picture came from the wreck of the Marie Rose. It lay under the sea for over 400 years.
- Many underwater craft (submersibles) are made of titanium because it is so strong and corrosion resistant.

20

10.2 Beating corrosion using alloys

You can't use an unreactive metal to solve every corrosion problem. For one thing, most unreactive metals are expensive. For another, they are often unsuitable for the job. Even if gold was cheap and plentiful, it wouldn't be used for bridge building! It's far too soft and weak for that.

Most corrosion problems are solved using **metal alloys** – mixtures of two or more metals. Alloys are usually made by melting metals together, then allowing the molten mixture to cool and harden. By mixing different metals, it is possible to make alloys which behave in very different ways.

Some mixtures of metals produce alloys which stand up to chemical attack. These are called **corrosion-resistant alloys**. Stainless steel is perhaps the best known of the corrosion resistant alloys. It is one of the family of **steels** – alloys of iron with other metals. In stainless steel, the iron is alloyed with 10–25% chromium. This converts the iron, which rusts easily and is quite brittle, into a stronger, rust-resistant metal.

Other corrosion resistant alloys are given in the table below. They all behave differently – each has been specifically designed to solve a particular corrosion problem and each has a particular use. More of that in the questions!

Stainless-steel jewellery

Name of alloy	Metals in alloy	What is the alloy like?
Corronel	Nickel, molybdenum, iron	Very resistant to acid
Solder	Lead, tin	Low melting point, good conductor
Burmabright 5	Aluminium, magnesium, with other metals	Light, very resistant to salt water
Amalcap	Mercury, silver, tin	Can be mixed cold, sets hard in 24 hours
Nimonic	Nickel, chromium	Is not affected by corrosion, even at high temperatures

Concorde—built with aluminium alloy

1 What is: a) an alloy   b) a corrosion-resistant alloy
   c) a steel? ▲
2 a) In what ways is stainless steel better than iron? ▲
   b) What is stainless steel used for? Give examples.
3 a) How is an alloy normally made? ▲
   b) Suggest why Amalcap alloy can be made without heating.
4 The alloys in the table were specially designed for making: *jet engines; the upper decks of ships; fillings for teeth; connections between electrical wires; reaction vessels in the chemical industry.* Which alloy was designed for which job? Match up each alloy with its use, explaining your choice.
5 **Try to find out:** about the alloys brass and bronze.

Did you know?

- By alloying iron with manganese, it is possible to make an alloy which is strong enough to break rocks.
- Many aeroplanes are built from aluminium alloys. They are light and corrosion resistant.

10.3 Acids and alkalis

There are very many different chemicals – so many that it's impossible to study them all. To make things simpler, scientists put chemicals which behave the same way into groups or sets. You have already met the set of metals. **Acids** and **alkalis** are two more of these chemical sets.

You can find out if a substance is an acid, or an alkali, by dissolving it and adding universal indicator. The diagram shows you how to do this.

The indicator's colour changes if acid or alkali is added to it. And so, by matching the indicator colour to a colour chart, you can find out what kind of substance has been added.

When acids and alkalis are mixed, they cancel each other out. You can use indicator to show this also.

The experiment on the right shows what happens when acid is added to alkali with indicator in it. At the beginning, before any acid has been added, the indicator is violet. Each time acid is added, it cancels out some of the alkali. This makes the indicator colour gradually change. Eventually the indicator goes green. Then the solution is **neutral** (neither acid nor alkali). Exactly the right amount of acid has been added to cancel out or **neutralise** all the alkali.

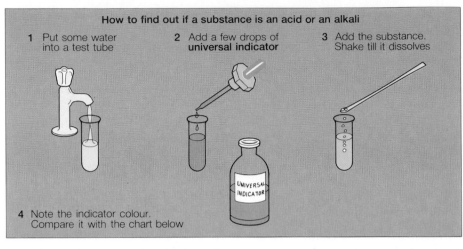

How to find out if a substance is an acid or an alkali

1 Put some water into a test tube
2 Add a few drops of **universal indicator**
3 Add the substance. Shake till it dissolves
4 Note the indicator colour. Compare it with the chart below

Universal indicator colour chart

Very acid substances	Slightly acid substances	Neutral substances	Slightly alkaline substances	Very alkaline substances
hydrochloric and sulphuric acid	vinegar fruit juice lemonade	water, sugar and salt	ammonia solution, detergent, baking soda, indigestion mixture	sodium hydroxide (caustic soda)

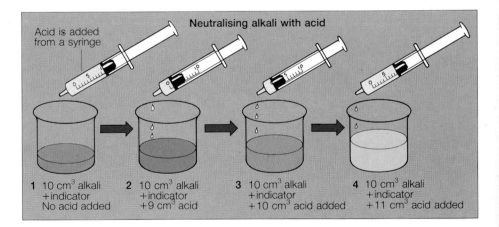

Neutralising alkali with acid

Acid is added from a syringe

1 10 cm³ alkali +indicator No acid added
2 10 cm³ alkali +indicator +9 cm³ acid
3 10 cm³ alkali +indicator +10 cm³ acid added
4 10 cm³ alkali +indicator +11 cm³ acid added

1   What is universal indicator used for? ▲
2   a) How would you find out if soluble aspirin is an acid or alkali?
    b) Aspirin turns indicator pink. What does this show?
3   a) What is a neutral substance? ▲
    b) Acid can neutralise alkali. What does this mean? ▲
4   a) Why does the indicator colour change in the neutralisation experiment above?
    b) Why is the indicator violet in 1, green in 3, pink in 4?
5   **Try to find out:** what the word 'acid' means.

Did you know?

● The dye in red cabbage can act as an indicator.

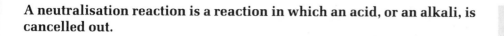
A neutralisation reaction is a reaction in which an acid, or an alkali, is cancelled out.

Neutralisation can be useful

Sometimes, your stomach makes too much acid. This can cause you to suffer from indigestion. Indigestion tablets help to cure the pain. They contain chemicals which cancel out the acid.

If you spill battery acid, you can cancel it out by adding a slightly alkaline substance like washing soda. But you should add lots of water first.

Tooth decay is caused by acids made in the mouth. Toothpastes contain chemicals that neutralize these acids and help stop tooth decay. That is why it is important to brush your teeth regularly.

Neutralisation produces salts.
When an alkali is exactly neutralised by an acid, a compound called a **salt** is made. You can't see anything happen when you mix the acid and alkali. (The acid and alkali are colourless solutions. The salt is colourless, too.) But when you evaporate off the water, the solid salt is left.

The most commonly made salt in the science lab is **sodium chloride**. It is made by mixing **sodium** hydroxide and hydro**chloric** acid. But there are many other salts, made by mixing different acids and alkalis. Some of these salts are useful as fertilisers, weedkillers, and drugs.

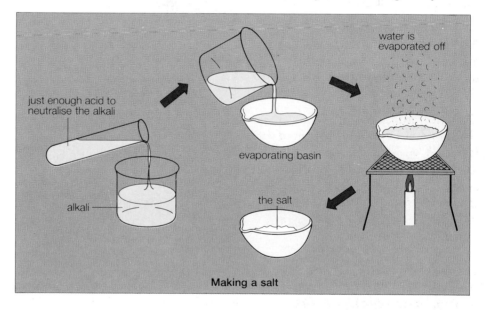

just enough acid to neutralise the alkali

alkali

water is evaporated off

evaporating basin

the salt

Making a salt

1   What is a neutralisation reaction? ▲
2   Why can you use: a) vinegar to treat a wasp sting   b) ammonia to treat a bee sting? ▲
3   a) Is baking soda solution acid or alkali (see p. 22)?
    b) Could you use baking soda solution to treat indigestion and a nettle sting? Explain your answer.
4   How would you make: a) a salt   b) a sample of sodium chloride? ▲
5   **Try to find out:** which salts are in bath salts and toothpaste.

Did you know?
● When you are stung by a nettle, hairs on the nettle leaf inject acid into your skin.
● Some types of ant defend themselves by squirting acid at their enemies. Up to ⅕ of the ant's body weight is made of acid.

10.3 The pH scale and soil testing

The pH scale. The pH scale measures how acidic (or how alkaline) a substance is. Most of the substances which you handle in the lab have pH values between 1 and 14. A substance with pH 1 is very acidic. A substance with pH 14 is very alkaline. A substance with pH 7 is neutral.

Soil testing. When the Smart family arrived at their new home, Fred looked at the garden, thought of his father, and shuddered. The garden was huge. His father was a gardening fanatic who did everything just as the book said. That meant lots of work for Fred!

It didn't take long for the first gardening job to come along. The very next morning, Mr. Smart was reading his gardening book in the section 'New Gardens'. 'To grow fruit and vegetables successfully', it said 'the soil pH must be correct. Each plant will grow best at a particular pH. Before planting anything in a new garden, make sure that you test the soil.' And so, that afternoon, Fred found himself out in the garden with a soil testing kit in one hand and a trowel in the other.

pH scale

pH	
14	
13	very alkaline
12	
11	slightly
10	alkaline
9	
8	
7	neutral
6	
5	slightly
4	acidic
3	
2	very acidic
1	

What Fred did. Fred tested soil from different parts of the garden, following the kit's instructions. These said:
1 Take several small samples of soil from different parts of the plot.
2 Dry the soil, then mix it.
3 Shake some soil with indicator.
4 Match the indicator colour to the pH chart.

What he found out. Fred drew a plan of the garden and put his results on it. He shaded each plot with the indicator colour given in the test and left his father to work out what his results meant.

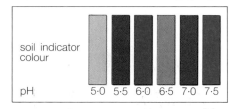

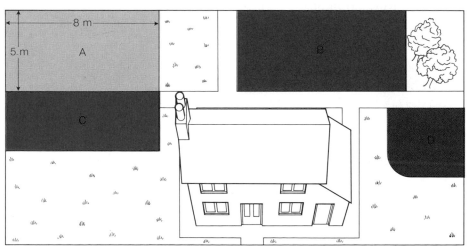

1 What is the pH scale used to measure? ▲
2 What can you tell about a substance which has: a) pH 1 b) pH 14? ▲
3 Suggest a pH value for: a) soap solution b) salt water c) battery acid (sulphuric acid) d) fruit juice. (p. 22 will help.)
4 Explain why: a) garden soils gradually become acidic b) lime is spread out on acid soil. ▲
About the Smart's garden:
5 Potatoes grow best at pH 5.5, turnips at pH 6.0. Use Fred's coloured plan to decide which plot should be used to grow each.
6 One of the plots had recently been limed. Suggest which.
7 Mr Smart asked Fred to lime plot A so that he could grow cabbages. They grow best at pH 6.5. Find: a) the area of the plot b) by how much the pH had to rise c) how much lime Fred had to use (150 g of lime raises the pH of 1 m² of soil by 0.5).

Did you know?

• Garden soil is likely to become more acidic as time passes. For one thing, the rain which falls is slightly acid. Many fertilisers are also acidic.
• Lime is spread on soils which are too acid. The lime cancels out the acid.

10.3 Acid tests

1 Crafty Claude was in custody, accused of stealing diesel oil from a farmer's tractor shed. The police claimed that powder found on his shoes was important evidence. They said that it was lime from the shed floor.

Claude said he was innocent. He claimed that he had been walking by the sea, and that the white powder was salt.

How could universal indicator have proved who was telling the truth?

2 Homes, factories and power stations which burn fuel, produce acid fumes into the atmosphere. These fumes do lots of damage. They dissolve in water in the atmosphere, making 'acid rain'. This acid rain causes widespread corrosion and poisons lakes, rivers, and forests.

The map opposite shows an industrial area, and the average pH of the rain which falls on the countryside around it.
a) Where does the most acidic rain fall?

b) Suggest why the rain is more acidic at B than at A.

c) Although the rain is more acidic at B, the lake there is less acidic than the lake at A. That's because the rocks at B are mostly limestone, while the rocks at A are granite. What does this suggest about limestone and granite?

3 Vinegar is made up of **acetic acid** dissolved in water. A food inspector was checking the quality of a large bottle of vinegar. He took 50 cm³ of the vinegar and cancelled out the acid in it using sodium hydroxide solution.

He knew that: 10 cm³ of his sodium hydroxide cancelled out 1 g of acetic acid.

He found that: 40 cm³ of his sodium hydroxide solution cancelled out the acetic acid in 50 cm³ vinegar.

The company's label on the bottle said 'This bottle contains 100 g of acetic acid in 1000 cm³ vinegar.'

Work out: a) how many grams of acetic acid were in the 50 cm³ sample of vinegar tested by the inspector b) how many grams were in 1 litre of the vinegar. Then suggest what the inspector would have said to the company!

sodium hydroxide solution

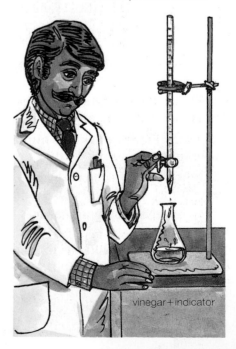

vinegar + indicator

25

The senses

Reggie is a robot. He has a computer for a brain.

The scientists who designed and built Reggie had lots of problems to solve. The biggest one of all was finding a way to control how he moved. After lots of thought, they designed a control system using the computer and a pair of microphones. They fixed the microphones to the side of his head. Then they wired them up so that electrical signals could be sent from the microphones to the computer. Lastly, they programmed the computer to make Reggie:

- go forward when they blew one blast on a whistle
- stop when they blew two blasts
- turn round when they blew three blasts.

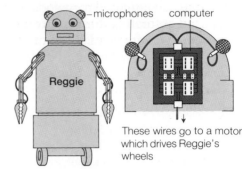

Reggie . . . and the inside of his head

At first, everything went well. Reggie moved round exactly as he was supposed to. But then, one day, disaster struck.

It was the ice-cream seller, Ernie's fault. He stopped his van in the street outside and blew one long blast on his whistle to let everyone know he was there. And Reggie moved forward . . . straight through the window . . . on the fifth floor!

It took scientists a year to make a new Reggie out of the mangled body of the old one. They had to make sure that he would never again have the same kind of accident! And so they redesigned him. Reggie Mark II has photocells fitted to the front of his head and touch pads to his hands. When he walks close to the window, the bright light makes the photocells send electrical signals to the computer. When he touches something the pad cells also send signals. And when the computer receives the signals, it makes him stop.

Reggie hasn't had another accident, but that's not surprising. Like a human, he has 'senses' to help him move around.

His photocells do the same job as human eyes. The eyes send signals to the brain when light enters them.

His microphones work like human ears. The ears also send signals to the brain – when sound enters them.

His touch pads do one of the jobs carried out by human skin. When something touches the skin, it send signals to the brain.

His wires are like human nerves. The nerves carry signals to and from the brain.

His computer acts like the human brain. The brain makes sense of the signals sent to it, then sends out instructions to make the body work.

But the human senses are far more complicated, and work far better than the most intelligent robot. You will soon find that out as you read through this section.

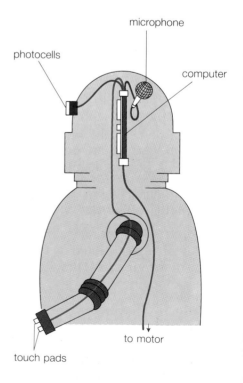

Reggie, Mark II

11.1 Mostly about light

No one can see anything in a completely darkened room. There has to be light before you can see! You see when light enters your eye. And so you have to know something about light to understand how the eye works.

What is light?

Light is a kind of energy. It's a type of radiation (like heat radiation). Light energy is given off by the Sun, by electric light bulbs, by candles, and other **light sources.**

Light energy travels from one place to another. The light travels in straight lines. (You can see that from the pictures of the Sun.) It's often useful to think of the light travelling as light rays. These rays are like tiny beams of light, travelling through the air at 300 000 km each second.

Shadows and reflections

Because light travels in straight lines, shadows form. **A shadow is an area behind an object where no light from a light source reaches.** When the Sun shines on an object, like a tree, the Sun's rays cannot bend round corners to reach the area behind it. That's why the area behind the tree is dark, and that's what makes the shadow.

Light rays can be bounced off, or **reflected** off, an object. You see this page because rays of light are being reflected from it into your eye. Shiny surfaces, like mirrors, reflect light best. You can see from the photo how a mirror reflects the rays.

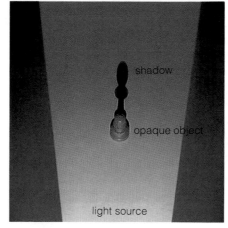

How shadows are formed
The rays from the torch can't bend round the object... and so a shadow forms behind it.

source of rays

mirror

Did you know?

- Pin-hole cameras took the earliest photographs. The film was fixed to the back of the camera.
- The Arabs and Chinese knew about lenses around 800 AD. 500 years after that, the first spectacles were made in Italy.

1 What is a light source? Give some examples.
2 Write down two facts about light rays.
3 What is meant by: a) a shadow b) reflecting light?
4 Why does a shadow form behind you when the Sun shines?
5 Mirrors and polished cars reflect light well. Walls don't. Why?
6 **Try to find out:** does the Moon give off light or reflect it?

11.1 Making pictures

Light can be used to make a picture on a screen. An easy way to do this is by using a pin-hole camera.

You can see a model pin-hole camera in the diagram. Light rays enter the box through the tine pin hole at the front to make a picture on the screen at the back. The hole is too small to let much light in.

You might think it would be easy to get a bright picture by making the hole bigger. This works but the picture becomes blurred. The big hole lets rays of light from one part of the lamp strike the screen at different places. **If you want to get a clear picture, all the rays from one part of the lamp must strike the screen at the same place.**

To get a bright, clear picture, you must use a wide hole and a **convex lens.** A lens is a specially shaped piece of glass which bends light rays. A convex lens bends light rays so that they come together. If you put the lens between the lamp and the hole, the picture becomes clearer. By moving the lens backwards and forwards, you can find a position which gives you a very clear picture. The picture is then **in focus**. The lens bends the rays of light which enter the camera from one part of the lamp so that they all strike the screen at the same place.

1 a) Explain how you can make a picture using a pin-hole camera. ▲
   b) Why is the picture not very bright? ▲
2 What does a convex lens do to light rays? ▲
3 **Try to find out:** how lenses are made.

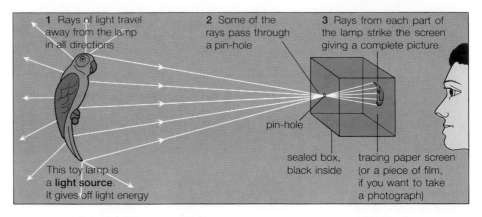

1 Rays of light travel away from the lamp in all directions

This toy lamp is a **light source**. It gives off light energy

2 Some of the rays pass through a pin-hole

3 Rays from each part of the lamp strike the screen giving a complete picture

pin-hole

sealed box, black inside

tracing paper screen (or a piece of film, if you want to take a photograph)

a small pin-hole sharp, but not bright enough

a large pin-hole bright, but blurred

a large pin-hole with a camera lens bright and sharp

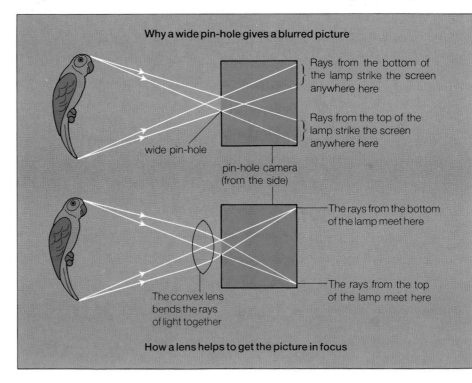

Why a wide pin-hole gives a blurred picture

Rays from the bottom of the lamp strike the screen anywhere here

Rays from the top of the lamp strike the screen anywhere here

wide pin-hole

pin-hole camera (from the side)

The rays from the bottom of the lamp meet here

The rays from the top of the lamp meet here

The convex lens bends the rays of light together

How a lens helps to get the picture in focus

11.1   Still pictures

To take a photograph, you need a film. A photographic film is a roll of thin, clear plastic coated with **light sensitive chemicals** (chemicals which are changed by light).

To take a photograph, you need a camera. The camera stores the film in darkness until you want to take a photograph. Then it lets light onto the film to make a 'picture' on the chemicals.

A suggestion

Why not look at a camera as you work through this page?

The camera . . .

The simplest cameras consist of:
a **light-proof box**, completely dark inside, which stores the film.
an aperture, a hole which lets light into the camera, to strike the film.
a shutter, which makes sure that light only strikes the film when you want to take a photograph. It covers over the aperture until you press the shutter release, then it opens and closes letting light in.
a lens, which focuses rays of light to give a clear photograph.

. . . and its controls

All but the simplest cameras have controls to help them take better photographs. These two controls are found on most cameras.
The distance control   Turning this control moves the lens in or out to get the picture in focus. (The closer the object to be photographed, the further the lens must be from the film.)
The aperture control   Turning this control makes the camera's **diaphragm** open or close. This changes the size of the aperture and controls the amount of light striking the film.

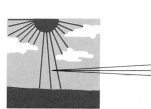

Taking a photograph

1 When you press the shutter release button

2 The shutter opens to let light into the camera for a short time

3 The light strikes the film, changing the chemicals

Different parts of the film are affected by different amounts of light. This gives the picture

The aperture

diaphragm

aperture

The size of the aperture is described by the f-number. In bright light, a small aperture, with a large f-number, should be used

The aperture settings

| f 16 | 11 | 8 | 5.6 | 4 |

bright sun hazy sun cloudy flash

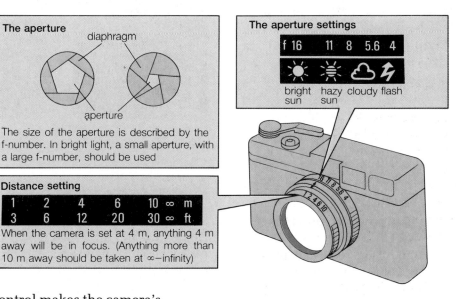

Distance setting

| 1 | 2 | 4 | 6 | 10 | ∞ | m |
| 3 | 6 | 12 | 20 | 30 | ∞ | ft |

When the camera is set at 4 m, anything 4 m away will be in focus. (Anything more than 10 m away should be taken at ∞ — infinity)

1   Why must a camera be a light proof box? ▲
2   Why must the camera have: a) an aperture   b) a shutter   c) a lens? ▲
3   Describe: a) photographic film   b) how a photograph is made. ▲
4   What happens when you turn: a) the distance control   b) the aperture control? Why is it necessary to turn these controls? ▲
5   When would you set the aperture control: to   a) f5.6   b) f16, and why?
6   Which distance and aperture settings would you use to photograph: a) a mountain scene in a thunderstorm   b) a rose on a sunny day?
7   **Try to find out:** how the earliest photographs were taken.

Did you know?

- Expensive cameras have **a shutter speed control**. It changes the length of time for which the shutter is open.
- The chemicals used to make the first photographs changed very slowly. Sometimes the shutter had to be open for hours!

Your eye – an energy changer

You probably know that your eye works when light enters into it. But did you know that the eye's job is to change the light energy into tiny electrical signals? The eye is a complicated energy changer.

The energy change takes place in the **retina** at the back of the eye. It contains millions of nerve cells. When light strikes a nerve cell an electrical signal is set up. This signal travels from the cell along a nerve to the brain.

So what happens when you see?

When you look at an object which is giving off light (like the toy lamp in the diagram), rays of light enter your eye. They pass through the **cornea**, **pupil**, and **lens**, then strike the retina. A 'light picture' is made on the retina. (It's very like the picture on the pin-hole camera screen.) The nerve cells which are affected by the light send signals to the brain. The picture which you 'see' is built up by the brain using information from these signals.

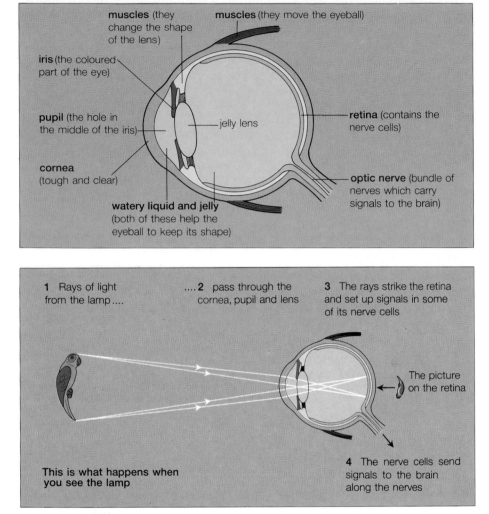

muscles (they change the shape of the lens)
muscles (they move the eyeball)
iris (the coloured part of the eye)
pupil (the hole in the middle of the iris)
jelly lens
retina (contains the nerve cells)
cornea (tough and clear)
optic nerve (bundle of nerves which carry signals to the brain)
watery liquid and jelly (both of these help the eyeball to keep its shape)

1 Rays of light from the lamp....
....2 pass through the cornea, pupil and lens
3 The rays strike the retina and set up signals in some of its nerve cells
The picture on the retina
This is what happens when you see the lamp
4 The nerve cells send signals to the brain along the nerves

The picture on the retina must be 'in focus' if you are going to see clearly. The cornea helps with this. It bends rays of light which pass through it. The lens also helps. It is made of a jelly-like substance. Round about it is a ring of muscles which make its shape change. By changing the thickness of the lens, the eye can focus light from far away and nearby objects.

The amount of light which strikes the retina must be controlled. The iris does this. It changes the size of the pupil making it small in bright light but wider in dim light. This makes sure that the best amount of light strikes the retina at all times.

1 What energy change takes place in your eye? ▲
2 You see a picture when light enters your eyes. What part does:
   a) the retina   b) the optic nerve   c) the brain play in building up the picture? ▲
3 What is the job of: a) the iris   b) the lens? ▲
4 Why is it important that the jelly lens can change shape?
5 Which of the eyes in the picture is looking at dim light?
6 **Try to find out:** some reasons why people wear spectacles.

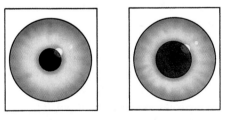

Did you know?

- There are around 120 million nerve cells in the retina.
- The lens in your eye is thinner when you are looking into the distance, and fatter when you are looking at something close-up.

11.2 Two eyes are better than one

Going further

If you ever have to wear an eyepatch over your eye, you will realise just how useful it is to have two eyes.

With two eyes, you can see more of what is round about you
With only one eye open, you can only see a little of what is round about you (unless you turn your head). With both eyes open, you can see far more – almost twice as much.

With two eyes, you can judge distances
If you try to catch a ball with one eye open, you will probably miss it. The information sent by one eye allows the brain to work out the direction in which the ball is coming. But it does not allow the brain to work out exactly how far away the ball is. The brain needs information from two eyes to do that.

Perhaps you have noticed that some animals and birds have eyes at the side of the head while others have eyes at the front.

When you look at the box

your left eye sees this   your right eye sees this

When you put the lid on the box, your brain uses information from both 'pictures' to judge how far away the box is

Owls (and other birds of prey) can judge distances well, because their eyes are at the front. This helps them to catch the prey they feed on. Monkeys also have eyes at the front. This allows them to judge distances for swinging through the trees.

A rabbit's eyes are at the side of its head. This allows the rabbit to see most of what is around without moving its head. A rabbit is more likely to survive if it can spot its enemies early.

A chameleon has the best of both worlds! It can move its eyes to the side when looking for prey, then move them to the front when it is ready to strike.

1 Give reasons why 'two eyes are better than one'. ▲
2 Why is it difficult to catch a ball with one eye closed? ▲
3 Why is it useful for: a) eagles and monkeys to have eyes at the front of the head b) rabbits to have eyes at the side? ▲
4 Why does a chameleon move its eyes to the front when it spots its prey?
5 Why do racehorses sometimes wear blinkers?
6 **Try to find out:** which of the following animals have eyes at the front of the head, and which at the side:
sheep fox deer horse dog lion antelope eagle
Then find the pattern in your answers.

Did you know?

• A chameleon can have one eye looking forwards and one eye looking backwards at the same time.
• An eagle's eyesight is good enough for it to be able to spot a hare 3 km away.

31

11.2 Moving pictures

You don't have to be able to drive to put a car into a garage! All you need is a piece of stiff card and a knitting needle. It's an old trick, but good. And it does teach you something about how your eye works.

To do the trick, you have to draw a picture of a car on one side of the card, and a picture of a garage on the other. If you fix the card firmly to the knitting needle, you can spin it rapidly. As you watch, the car appears to go into the garage.

The trick works because: **the picture on the retina of the eye lasts for about 0.1 second before fading.**

When light from the car picture strikes the retina the nerve cells send signals to the brain. These signals last for about 0.1 second. If you spin the card rapidly, light from the garage picture will strike the retina before 0.1 second has passed. And so the nerve cells send signals from the garage picture to the brain before signals from the car picture have faded. This confuses the brain. It cannot separate the two pictures. It 'sees' the car and the garage at the same time. In other words it 'sees' the car in the garage.

This effect is used by makers of 'movie pictures'. A movie camera is like an ordinary camera in many ways, but there is one main difference. To make a movie, film is run through the camera and the shutter is opened and shut automatically 24 times a second. In this way, 24 separate photographs or **frames** are taken on the film each second, each frame being slightly different from the one before it.

When the developed film is run through a projector, pictures are flashed on the screen at a rate of 24 each second. This is so rapid that light from a second picture strikes the retina before the picture of the first has faded. This means that the brain receives signals from different pictures at the same time. It 'sees' the rapid series of still pictures as a smoothly moving picture.

Did you know?

- Making a cartoon film is hard work for the artists. They have to draw 24 pictures for each second of film!
- Old fashioned movie cameras were turned by hand. They used a speed of 16 frames per second. That's why old films are jerky.

1 How long does the picture normally last on the retina? ▲
2 Why is the brain confused if different pictures are flashed on to the retina within 0.1 second of each other? ▲
3 Why, in the trick, does the brain 'see' the car in the garage? ▲
4 What is the main difference between a still camera and a movie camera? ▲
5 Why does the brain 'see' a moving picture when still pictures are flashed onto a screen using a movie camera? ▲
6 a) How many frames are there in a 10 minute cartoon?
   b) How long does the cartoon cowboy take to draw his gun?
7 **Try to find out:** more about making cartoon films.

11.3 Making sound – making vibrations

Sound, like light, is a kind of energy. Sound energy, like light energy, can travel from one place to another. Your ears can detect this energy. That's one way you can find out what is going on around you.

How sounds are made

In a rock band, there are usually several different types of instrument. The different instruments make different sounds. Yet all of the sounds are produced in the same kind of way. Like all sounds, **they are made by things which are vibrating backwards and forwards.**

The guitarists pluck the guitar strings and make them vibrate. The saxophonist blows into the saxophone, making the air inside vibrate. The drummer strikes the drum and makes the drum skin vibrate.

Even the singer depends on vibrations! Her sound is made by her **vocal cords**. These are flaps of skin inside her wind pipe. When she sings, she blows air over the flaps and makes them vibrate.

How sound travels – music to your ears!

Sound can travel through the air. You don't have to put your ear on the drumskin to hear the drum beat!
The sound is carried to your ears by vibrating air molecules.

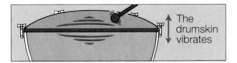

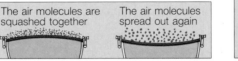

1 When the drummer strikes the drum, the drumskin vibrates rapidly up and down. The size of these vibrations is called the **amplitude**.

2 The vibrating drumskin makes the molecules vibrate backwards and forwards. These molecules affect the molecules next to them. The sound spreads out.

3 Within a fraction of a second, all the air molecules in the theatre will be vibrating. You hear sound when the air inside your ear starts to vibrate.

1 What do sound and light have in common? ▲
2 How are sounds made? ▲
3 What makes the sound when you are playing: a) a guitar
   b) a drum   c) a saxophone? ▲
4 List some musical instruments which are played by striking them.
5 Why is a sound made when: a) you run your fingers along the teeth of a comb   b) you slam a door   c) wind blows through telephone wires?
6 What are vocal cords? How do they allow you to sing and talk? ▲
7 Explain how sound is carried from a drum to your ears. ▲
8 **Try to find out:** how a piano's soft pedal works.

Did you know?

● Light travels much faster than sound. The speed of light in air is 300 000 km each second!

Your ear is another of your body's energy changers. Its job is to change sound energy into electrical signals which are sent to the brain.

When sound vibrations from a drum, or any other vibrating object, reach your ear, they make the air inside vibrate. This is what happens next:

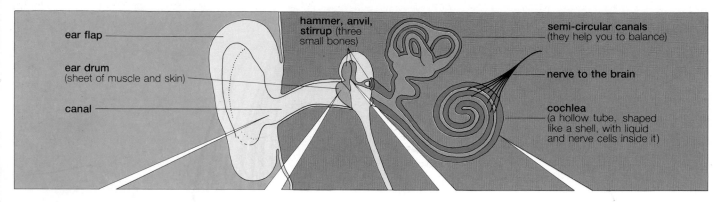

1 The air inside the canal begins to vibrate. This makes . . .

2 . . . the ear drum start to vibrate. This makes . . .

3 . . . the bones start to vibrate. This makes . . .

4 . . . the liquid in the cochlea vibrate. This sets up signals in the nerve cells.

The electrical signals travel from the nerve cells, along a nerve, to the brain. Different nerve cells are affected by different sounds. When a band is playing, the pattern of signals which reaches the brain is constantly changing. The sounds you hear depend on the information in the signals reaching the brain.

When the ear stops working

Your ear is very delicate and can be easily damaged. If any one part of it is damaged, you could go deaf.

There are several different causes of deafness. Some people are deaf because sound vibrations can't travel through their ears. This kind of deafness may be caused by an accident which damages the ear drum or the bones, or simply by wax blocking the ears. Other people are deaf because their nerve cells are damaged. Old people are most affected by this. In old age, their nerve cells don't work as well as they did before. But younger people are beginning to suffer from this kind of deafness. Nerve cells are damaged by loud noise!

Cutting down trees is a noisy business. That's why the forester wears ear protectors

1 What energy change takes place in your ear? ▲
2 What are: a) the ear drum b) the hammer c) the cochlea? ▲
3 How are sound vibrations carried through the ear to the nerve cells? ▲
4 a) Write down two causes of deafness. ▲
  b) How does a doctor cure deafness caused by wax in the ears?
5 Why are noisy discos bad for your hearing?
6 **Try to find out:** the job done by the semi-circular canals.

Did you know?

- The stirrup bone is the smallest bone in the body. It is only 3 mm long.
- If you ever have to work in a noisy factory, you should wear ear protectors. They could prevent you from going deaf in old age.

Sound can travel through gases. Most of the sounds which you hear travel to your ears through the air.

Sound can travel through liquids. You can hear sounds when you are swimming underwater. Dolphins 'speak' to each other underwater by sending out high pitched squeaks and clicks.

Sound can travel through solids. When you press your ear to the wall, you can hear sounds from the next room.

Sound can travel through solids

but not through a vacuum

But sounds can't travel through a vacuum. Sound vibrations need atoms or molecules to carry them from one place to another. A vacuum is an empty space with no particles in it, and so it can't carry sound.

How fast and far does sound travel?

If you are standing near the finish line of a 100 m race when the starter fires his gun, the sound will take only ⅓ second to reach you. Sound travels through air at a speed of 330 metres each second.

But sound can travel even more quickly through some substances. Hard solids carry sound fastest. In 1 second, sound travels 5100 m through iron and aluminium, 3600 m through brick, 6000 m through glass. Sound also travels rapidly through liquids – 1400 m each second through water.

If you are standing a long way from the starter's gun, you may not hear it fire. Sound vibrations lose energy as they travel. Eventually they become so weak that you can't hear the sound.

How far the vibrations travel before this happens depends on what they are travelling through. For example, sound can travel much further through hard solids than through soft materials with air trapped in them. That's why thick curtains can be very useful if you live near a noisy motorway. The curtain material cuts down a lot of sound which would otherwise enter the room.

I can't hear what you're saying!

Did you know?

- There is silence on the Moon. It has no atmosphere to carry sound.
- Double glazing can cut down sound – but only if there is a large gap (about 100 mm) between the two panes of glass.

1   What evidence is there that sound travels through:
    a) solids   b) liquids   c) gases? ▲
2   Why can't sound travel: a) in a vacuum   b) on the Moon? ▲
3   What kind of material carries sound: a) furthest   b) fastest? ▲
4   Why can thick curtains be useful if you live near a motorway? ▲
5   Suggest: a) why the cartoon Indian has his ear to the ground
    b) why Fred's alarm clock isn't disturbing his long lie in.
6   Use information above to complete the 'speed of sound' bar chart.

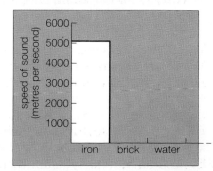

11.3　High and low

Frequency and pitch

Do you have a high pitched voice?
Whether your voice is high pitched or low pitched depends on how rapidly your vocal cords vibrate when you speak. The faster the cords vibrate, the higher will be the pitch of your voice.

The pitch of a sound (how high or low the sound is) depends on how rapidly the 'sound producer' vibrates.

The frequency of a sound is the number of sound vibrations set up in 1 second. It is measured in Hertz (Hz). When you pluck the top string of a guitar, it vibrates 660 times in 1 second. The frequency of the sound it makes is 660 Hz. When you pluck the bottom string, it vibrates only 165 times each second, giving a sound with frequency 165 Hz.

If you have ever played a guitar, you should now realise the connection between the frequency of a sound and its pitch. The top string makes a higher pitched sound than the bottom one.

The higher the frequency of a sound, the higher is its pitch.

Can you hear what they are saying?

You will probably be able to hear a wide range of sounds – from the low pitched note of a drum (frequency 20 Hz) to a very high pitched whistle (frequency 20 000 Hz). But many animals can hear, and make, a much wider range of sound than humans can. You can see information about a few of these animals in the diagram.

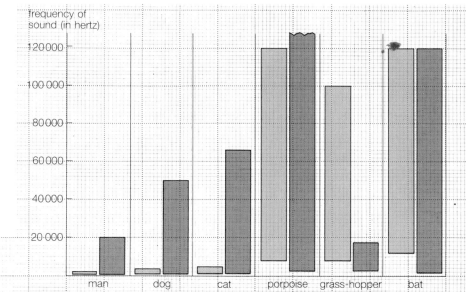

the frequencies of sound which the animal can hear

the frequencies of sound which the animal can make

1　What is meant by:　a) the pitch　b) the frequency of sound? ▲
2　The frequency of a drum note is 20 Hz. What does this tell you about the drumskin movement when it is struck?
3　Will your vocal cords be vibrating more rapidly when you give a high pitched scream or a low growl? Explain your answer. ▲
　Look at the diagram of 'sounds made' and 'sounds heard'.
4　Write down the frequencies of:　a) the highest sound made by a porpoise　b) the lowest note heard by a bat　c) the sounds heard by a dog.
5　Which animal (or animals):　a) hears the highest sound　b) make sounds they can't hear　c) make sounds you can't hear.
6　**Try to find out:** why you don't hear a dog whistle properly.

Did you know?

- Older people can't hear some high pitched sounds which you can.
- Porpoises 'talk' to each other using low frequency barks, moans and whistles, but send out high power, high frequency sounds to help them navigate.

11.4 Smell and taste

There are 'smell cells' in your nose and taste cells (called **taste buds**) in your tongue. When you are eating your favourite type of crisps, both sets of cells work together to give you the flavour you enjoy.

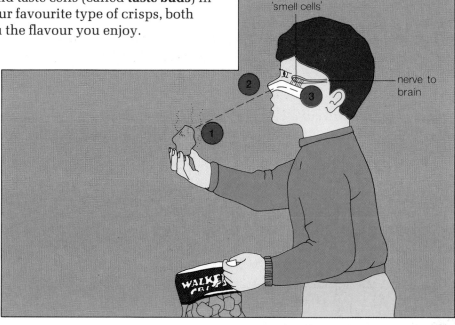

This is how your nose detects the smell from the crisps:

1 Chemicals in the crisps give them their smell. Some of these chemicals spread through the air.

2 When you breathe in, these chemicals enter your nose. There they dissolve in the liquid which moistens all of the inside of the nose.

3 The dissolved chemicals set up signals in the smell cells. These signals travel along nerves to the brain.

This is how your tongue detects the crisps' taste:

The surface of your tongue is rough. It is covered with tiny bumps which contain the taste buds. When you chew a crisp, the flavouring chemicals dissolve in the liquid on your tongue. This affects the taste buds which send signals to the brain.

There are four main kinds of taste: sweet, sour, salt, and bitter. As you can see from the diagram, different areas of the tongue are better at detecting different tastes.

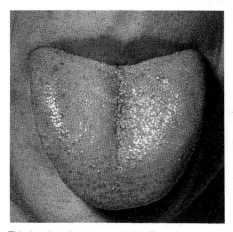

This is what the tongue looks like, close-up

Did you know?

- You probably have about 20 000 000 smell cells but only 9000 taste buds.
- You can get used to a smell. After some time, your smell cells stop sending messages to the brain.

1 When you open a packet of crisps, your nose detects a smell almost immediately. Explain how this happens. ▲
2 What are taste buds? What is their job? ▲
3 Explain how your tongue detects a taste when you are licking an ice-cream.
4 Which part of your tongue (tip, side or back) is sending most messages to the brain when you are: a) eating an orange   b) licking a lollipop   c) eating salty crisps   d) tasting vinegar.
5 Some medicine is unpleasant to take. Why does it help to hold your nose when you are taking it.?
6 **Try to find out:** what a perfume blender does.

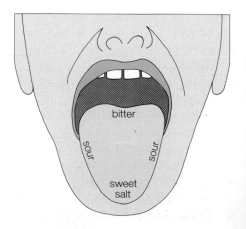

The skunk is one of the smelliest animals in nature. When another animal gets too close, the skunk squirts a very smelly liquid at it.

A dog has an excellent sense of smell. It can use scent to hunt a rabbit, or to track a man through a crowd.

The female gypsy moth attracts a mate by releasing special scents into the air.

Salmon may travel thousands of miles to return to the gravel beds where they were born to lay their eggs. They are thought to find their way across the oceans using light from the Sun and Moon, but as they swim round the coast, they recognise the correct river by its smell.

Deer depend on scent as well as sight for keeping out of danger. This deer is on the alert because it has smelt a human being. A deer stalker has to stay downwind of a deer if he is going to get close to it.

The fox uses scent for hunting. It can also use scent for finding its way about. It has special scent glands on its feet and these leave scent on the ground when the fox moves about. This allows the fox to retrace its steps.

1 Which of the animals in the photographs use scent or smells:
   a) to hunt their prey   b) for defence   c) to help them find their way about?
2 How is scent useful to: a) the gypsy moth   b) the fox?
3 How can scent put a fox in danger?
4 Suggest why mountain rescue teams use dogs on a rescue.
5 When could your sense of smell warn you of danger? (Give as many examples as you can.)
6 Write a few sentences about 'Smells I like' or 'Smells I dislike'.
7 **Try to find out:** about jobs done by police sniffer dogs.

Did you know?

- An Emperor moth can detect the scent of a female 11 km away. The female only carries 0.000 000 1 g of scent.
- A salmon can also detect tiny amounts of scent. One molecule of scent is all that is needed.

11.5 Information from the skin

It might surprise you to learn that your skin is as much of a sense organ as your eyes or ears. In fact, it contains millions of nerve endings and nerve cells which send information to the brain. These nerve cells and nerve endings do not lie right on the surface of the skin. (The outside layer of the skin is made up of dead skin cells.) They are found in the living layer underneath. This layer contains cells which:

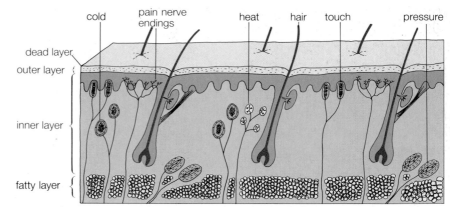

The skin's nerve cells

are affected by **pain**

are affected by **pressure**

send information about **touch**

send signals when the skin gets **hotter**

send signals when the skin gets **colder**

1 What kinds of information are supplied to the brain by the skin? ▲
2 The hot cells in your skin send signals to the brain when you touch a hot radiator. When will: a) your 'cold cells' b) your pressure cells c) your pain cells send signals? Give examples.
3 Explain why: a) you can cut the very outside layer of your skin without feeling pain b) it hurts to cut deeper.
4 Feeling pain isn't pleasant, but it can be useful. Why is it useful when you touch a hot kettle or step on glass?
5 **Try to find out:** why a layer of dead skin is a useful covering for the body.

Did you know?

- Curry powder affects the hot cells in your tongue. That's why curries seem to be hot.
- Some nerve endings send signals to the brain when the skin is touched very lightly. That's what happens when you are tickled.

The skin nerve cells and nerve endings are not evenly spread all over your body. For example, there are more 'hot cells' in your fingertips than in your mouth. There are far more pain nerve endings in your tongue, lips and fingertips than in your thigh.

A part of the skin which has a large number of nerve cells or nerve endings is described as **highly sensitive**. Your tongue is highly sensitive to pain. If you cut it, it really hurts! That's because the tongue has many pain nerve endings which can send 'pain signals' to the brain.

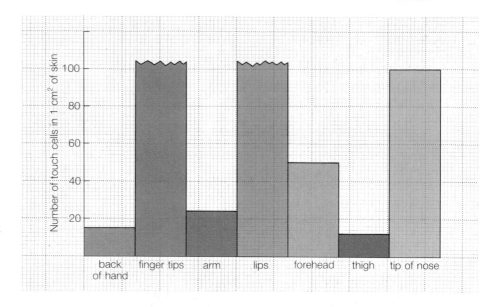

Some parts of the body are far more sensitive to touch than others. The bar chart above shows you that. There are large numbers of touch cells in the skin of your fingertips. Fingertips are highly sensitive, and that can be useful.

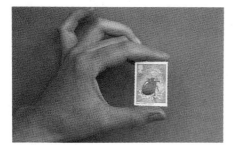

You can hold a delicate object safely with your fingertips. The fingertips send lots of signals to the brain. So the brain has lots of information which it can use to make the fingers hold the object lightly, without crushing it.

The Braille alphabet allows blind people to read using their fingertips. Each letter in the alphabet is a series of raised dots on the paper. A skilled Braille reader can read 60 words a minute by running his fingers over the paper.

You can use your fingertips to find out how rough or smooth a surface is. When you are sanding a piece of wood, you can run your fingertips across its surface to check whether it is smooth enough.

1 What is meant by a 'highly sensitive part of your skin'? ▲
2 Why is a small cut on your tongue very painful? ▲
3 Why can you safely hold delicate objects with your fingertips? ▲
4 Explain what happens when a blind person reads by Braille. ▲
5 Using the bar chart: a) name three parts of the body which are sensitive to touch b) write down the number of touch cells in 1 cm² of forehead skin c) work out the number of touch cells on the back of your hand. (You can take the area of skin to be 40 cm².)
6 Why can you drink coffee which burns your fingers?
7 **Try to find out:** more about Braille.

Did you know?

- Louis Braille was a Frenchman who was blinded by an accident at the age of three. He invented the Braille language when only 15.
- Music can now be written in Braille.

11.6   What happens to the signals?

The body has a control centre. It's called the **brain**. It's made of over 1 kilogram of soft grey and pink material and it lies inside the skull.

Nerves run between the brain and the rest of the body. They carry signals to and from the brain. A large number of the nerves, grouped together, make up the **spinal cord**. The cord runs from the brain down through a tunnel inside the backbone. The backbone gives important protection to the nerves.

Your brain and nerves make up your **nervous system**. It controls much of what goes on in your body. At all times, millions of signals flow to and from the brain along the nerves. Many of the signals flowing to the brain carry information from the eyes, ears, nose, tongue and skin. Most of the signals flowing from the brain go to the muscles.

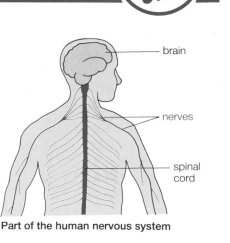

Part of the human nervous system

The brain's job is to sort out, then send out signals. It:

- collects signals sent to it
- makes sense of the information they contain
- decides what to do
- sends signals to the muscles to make them take action.

Choosing a cake is a **'thinking action'**. Your brain decides which cake your hand should pick up.

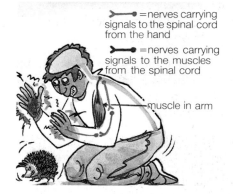

Reflex actions

Some of your body actions take place without your brain being involved. These automatic actions are called **reflexes**. If your hand touches a hot cooker, you will pull it away immediately. That's a reflex action. Coughing and sneezing are other reflexes. They take place automatically when harmful substances get into the nose and lungs.

A thinking action
Signals from eye to brain
'There are lots of different cakes on the plate.'
Brain decides 'The chocolate cake looks best.'
Signals from brain to muscles in arm and hand 'Pick up the chocolate one'

A reflex action
Signals from the pain cells travel to the spinal cord. There they set up signals in the nerve running to the arm muscles. The muscles jerk the hand away.

The pain cells' signals travel on to the brain. You feel the pain *after* your arm has moved.

1 What is the nervous system made up of? ▲
2 What job is done by: a) the nerves   b) the brain? ▲
3 What protects: a) your brain   b) your spinal cord? ▲
4 A person who breaks his backbone is often paralysed in the lower part of his body. Suggest why.
5 Switching off a boiling kettle is a 'thinking action'. Using the model above, explain what is going on in your nervous system.
6 a) Coughing and sneezing are reflex actions. What does this mean? ▲
   b) These reflexes (like others) can help to protect you. Explain how.
7 **Try to find out:** about other reflexes in your body.

Did you know?

- The brain can deal with 50 million signals each second.
- The brain is easily damaged but it is well protected by the skull and by a cushion of liquid which surrounds it.

11.6 Learning is important

There is a huge amount of information stored up in your brain. Every time you learn something new, a piece of information is added to your memory store. This learned information is very important. Your brain uses it, along with the information from the senses, before sending out instructions to the muscles.

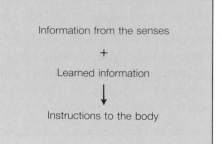

Information from the senses

+

Learned information

↓

Instructions to the body

Using learned information

Learning is all about storing up information. While you are being taught to do something, lots of new information is added to your memory store. Then, when the time comes for you to put the learning into practice, your brain can use this information in deciding what to do. Your actions will depend on what you have learned.

If you had to deal with these emergencies, suggest:
a) what you would probably do if you had no emergency training.
b) what is the correct thing to do. (If you don't know, find out!)

If you **have** been given emergency training, you will have been taught **exactly** what to do in the case of electric shock, or fire, or an accident on ice. Your brain simply has to call up the correct information and use it.

But your brain can do better than this. It can use learned information to solve problems which you have never met before. When you are faced with a new problem, your brain searches through the information in its memory store. It may contain information about a similar situation and this could help you to arrive at an answer.

How can Charlie cross the river?

Does your memory store contain enough 'learned information' to solve this problem? One piece of information which could help: polythene bags can be blown up like balloons.

11.6   The human's computer

In 1979, the author of a book about the brain wrote:

- To be as powerful as your brain, a computer would have to weigh 10 tonnes.
- There are as many nerve cells and connections in 1 gram of your brain as there are cables and connections in the world's telephone system.

Of course, computers have become much smaller since then, and the world's telephone system has become larger. But what the author wrote should give you some indication of how incredibly complicated the brain is. It contains more than 10 000 000 000 nerve cells. Each nerve cell can be connected to as many as 25 000 other cells. There are billions and billions of pathways which the brain signals can follow.

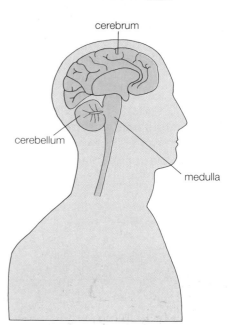

Different parts, different jobs

Although the brain is so complicated, scientists have gone some way towards understanding how it works. They know that different parts of the brain do different jobs.

The **cerebrum** is the part concerned with intelligence, memory and thought. It controls your body's **voluntary actions** (actions you control by thinking about them).

As you can see from the drawing, different areas of the cerebrum seem to be concerned with different activities which go on in the body. Not so much is known about the other areas. They are probably concerned with memory and thinking.

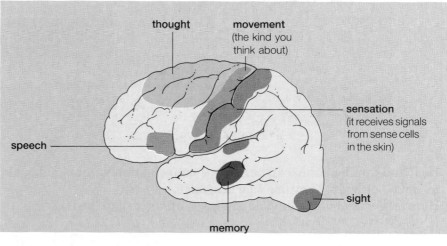

The **medulla** and **cerebellum** control many of your body's **involuntary actions** (actions which go all of the time without you thinking about them). The medulla controls heart rate, breathing, and swallowing. The cerebellum is concerned with balance and muscle control.

1   The cerebrum, medulla and cerebellum are parts of the brain. What does each part do? ▲
2   Which part of your brain: a) controls your pulse rate   b) controls your fingers when you sew   c) helps you to understand what your friends say   d) helps you balance on a tight rope?
3   a) What's the difference between a voluntary and an involuntary action? ▲   b) Which kind of action is digestion? Explain your choice.
4   Which areas of the cerebrum are working when you read out loud?
5   **Try to find out:** ways in which   a) your brain is better than a computer   b) a computer is better than your brain.

Did you know?

- The cerebrum is in two halves, left and right. The left half controls the right hand side of your body and the right half controls the left hand side.

43

11.7 Overcoming handicaps

You miss a lot if you have no sense of smell, or if you can't taste anything, but it isn't the end of the world! If you can't speak, or hear, or see, however, life is much more difficult.

In all sorts of ways, blind, deaf, or dumb people try to overcome their handicap.

Blind people read without seeing, using Braille.
In Braille, each letter is a pattern of raised dots. Here is the Braille alphabet, with part of a message for you to work out.

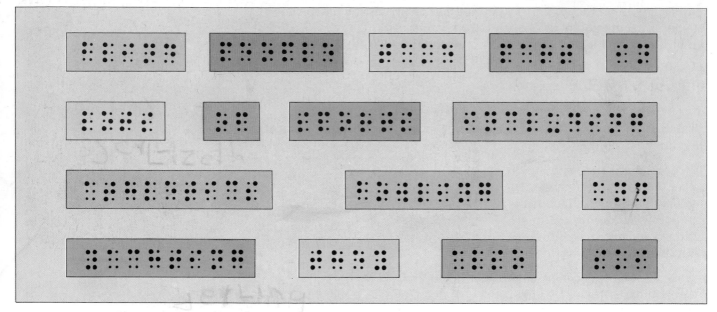

Deaf and dumb people can carry on a conversation without being able to speak, or hear. They use sign language.
Here is the 'Deaf and Dumb' language, with the rest of the message.

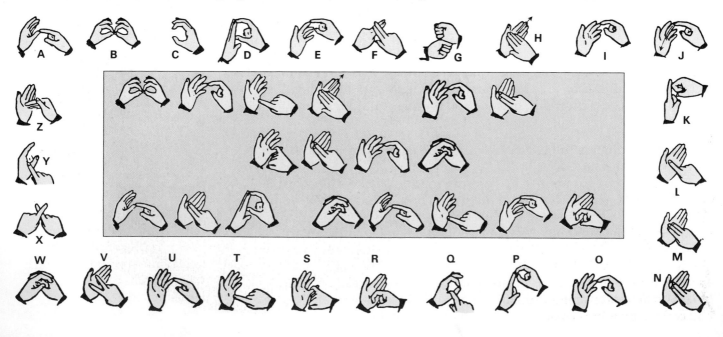

The Earth

This is what the outside of the Earth looks like. The photograph was taken by an astronaut, out in space.

This is what the Earth looks like inside. At least, that's what **geologists** believe. Geologists are scientists who study rocks. They think that the Earth is made of up three main layers – the **core**, the **mantle** and the **crust**.

The **crust** contains the least dense rock. It's much cooler than the rest of the Earth

The **core** is mostly made of iron. It is very dense and hot with temperatures between 2000 °C and 5000 °C. The inner core is solid. The outer core is red hot liquid rock

crust	mantle	outer core	inner core
5–90 km	3000 km	2255 km	1215 km

The **mantle** is cooler, around 1600 °C. The rocks in it are less dense than the rocks in the core

This section deals mainly with the crust. It's not difficult to see why. For one thing, it's the part that you live on. It also supplies many of the things you need for living, including:

building materials, **metals**, **fuels**, and **soil for growing crops**.

45

The Earth's crust contains hundreds of different rocks. Here are six which are found in Britain.

Granite is a very hard rock. It's not easily scratched. It sparkles in the sunshine because it has crystals in it. Some of these crystals are big enough to see.

Sandstone is a softer rock. It's made up of grains of sand. You can easily rub the grains off the rock, even with your finger nail!

Slate has crystals and is hard, but not as hard as granite. It is made up of layers which can be split off from each other. That's why pieces of slate are often flat.

Like granite, **basalt** is very hard, difficult to scratch and is made up of crystals. But the crystals in basalt are tiny.

Chalk is another soft rock. White grains come off when you rub it. That's why it can be used for writing on blackboards.

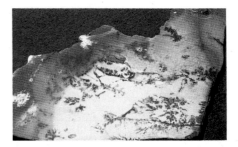

Marble is mostly white or grey, but may have streaks of colour. It contains small crystals. It is hard but can be scratched with a knife.

How can you pick out one rock from another?

There are simple clues which can help you to identify rocks. A rock's hardness and colour can give useful information (although some rocks, like granite, can have different colours). Any crystals, grains or layers in the rock can also help you to decide what it is. So can a bottle of acid! Some rocks, like marble and chalk, fizz when acid is dropped on them. But usually identifying rocks is a job for an expert. He will need more clues than those mentioned here!

1 From the rocks mentioned on this page, make sets of: a) hard rocks b) soft rocks c) rocks with crystals d) rocks with grains.
2 a) What information can help you to identify a rock? ▲
 b) Why can colour be confusing when you identify a rock? ▲
 c) A rock is soft, made of white grains and fizzes in acid. What could it be?
3 Give two reasons why rocks may be different from each other. ▲
4 Which should last longer, granite or sandstone pavements? Why?
5 Suggest what has made: a) chalk b) slate useful for writing.
6 **Try to find out:** where these rocks are found in Britain.

12.1 Rocks from the melting pot

Granite is one of a large number of rocks which have crystals in them. These crystals give a clue about how the rocks were formed.

Crystals of a solid can be made by:
1 *heating the solid until it melts to a liquid* then
2 *allowing the hot liquid to cool until it forms a solid again.*
Geologists believe that the crystals in granite, and in many other rocks, were made by the cooling of hot liquid rock called **magma**. Magma is formed under the Earth's surface in places where there is enough heat to melt the rock. Sometimes the molten rock pushes its way upwards. If it cools enough to solidify, it produces rocks with crystals in them. The slower the cooling, the bigger the crystals are.

Rock which is formed when hot liquid rock cools and hardens is called **igneous rock**. The diagram shows how different kinds of igneous rock can form.

Molten flowing lava

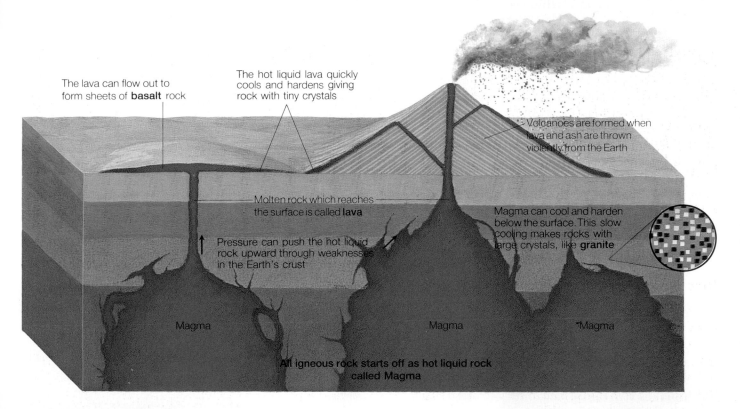

The lava can flow out to form sheets of **basalt** rock

The hot liquid lava quickly cools and hardens giving rock with tiny crystals

Volcanoes are formed when lava and ash are thrown violently from the Earth

Molten rock which reaches the surface is called **lava**

Magma can cool and harden below the surface. This slow cooling makes rocks with large crystals, like **granite**

Pressure can push the hot liquid rock upward through weaknesses in the Earth's crust

Magma

Magma

Magma

All igneous rock starts off as hot liquid rock called Magma

1 What is: a) igneous rock b) magma c) lava? ▲
2 Explain how: a) magma b) a volcano is formed. ▲
3 a) Why are granite and basalt called igneous rocks? ▲
   b) Look at the descriptions of granite and basalt on page 46. Write down what these rocks (and other igneous rocks) have in common.
4 a) How does the speed of cooling affect the size of the crystals in a rock? ▲
   b) Why are granite crystals much larger than the crystals in basalt?
5 **Try to find out:** what pumice stone is, how it was formed and why it has holes in it.

Did you know?

- Magmas from different regions can contain different chemicals. The lava flowing from Hawaiian volcanoes is different from that flowing from volcanoes in Iceland.

12.1 But how do geologists know?

The deepest hole ever drilled reached down over 10 km into the Earth, but even then, it was less than half way through the Earth's crust. That should raise a question in your mind. If no one has ever bored right through the Earth's crust, how do geologists know what the Earth is like inside?

Building up a picture of the inside of the Earth is a complicated business. Geologists have to take many different pieces of evidence and fit them together. Some of this evidence comes from volcanoes which bring molten rock to the surface from deep down. Most evidence, however, comes from **seismology**, the study of earthquakes.

An earthquake is produced when rocks inside the Earth suddenly slip. This causes huge movements of rock and sets up shock waves called **seismic waves** which travel through the Earth. The speed at which a seismic wave travels depends on what it is travelling through. By measuring the speeds at which seismic waves travel, geologists can build up a picture of what the Earth is like inside.

Edinburgh castle sits on top of the plug of an extinct volcano

The map opposite shows the Earth's main earthquake regions and its active volcanoes. You will notice that Britain does not lie in an 'active zone' now, but many millions of years ago, there were British volcanoes. Many parts of Britain have rocks made of lava.

You will notice, too, that earthquakes and volcanoes occur in the same areas. They are found where there are weaknesses, called **faults**, in the Earth's crust. Geologists believe that huge sections of the Earth's crust, called **plates**, are slowly drifting. When two plates meet, and push on each other, a fault is formed.

Geologists know where the faults are. They can predict where earthquakes will take place, but not exactly when.

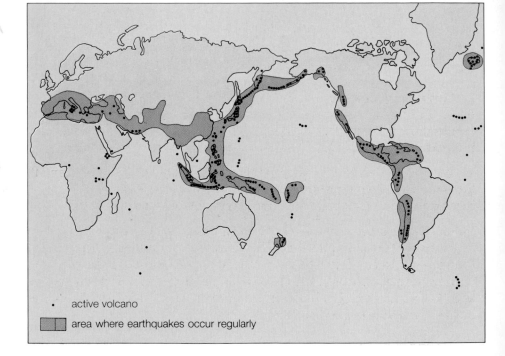

· active volcano

▨ area where earthquakes occur regularly

1 How is an earthquake produced? ▲
2 What is: a) seismology b) a seismic wave? ▲
3 Why do geologists find it useful to study seismic waves? ▲
4 What is a fault? Why are volcanoes produced at faults? ▲
5 Using the map, name some countries which are badly affected by earthquakes and volcanoes.
6 What evidence is there that Britain once had volcanoes? ▲
7 If you live near a railway, your house will be affected by seismic waves. Explain why.
8 **Try to find out:** how a seismograph works.

Did you know?

● Volcanic eruptions can be explosive. In 1815, an Indonesian volcano blew away the top 1 km of a mountain. In 1883, another Indonesian volcano, called Krakatoa, blew rocks 50 km into the air.

Mountains don't last for ever, even if they are made of granite! The rocks are attacked by all types of weather.

- Wind, rain, and snow beat down on the rock surface.
- Water seeps into cracks in the rock and then freezes.
- As the water turns to ice it expands, cracking the rock.
- Hot sunshine followed by cold night temperatures makes rocks expand and then contract. This also makes rocks crack.

All of this breaks bits off the rock. These bits can be large pebbles, pieces of gravel, small grains of sand, or even tinier grains of mud or clay. As the rock fragments are carried off by wind or water, the mountain is very, very slowly worn away. This wearing away is called **erosion**.

Glaciers grind down rocks. This river, running from a glacier, is white with rock fragments

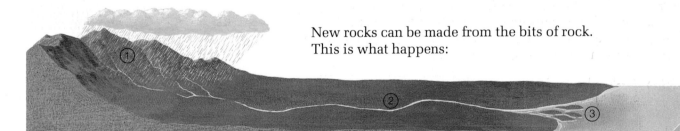

New rocks can be made from the bits of rock. This is what happens:

1 **Weathering** breaks down the rocks. The rock fragments are carried off by the wind or washed into streams and rivers.

2 The rivers carry the bits of rock down to the sea (or into a lake). Fast rivers can carry down bigger pebbles than slow ones.

3 The river water slows down as it reaches the sea. There, the sand, gravel and pebbles settles as sediment on the sea bed.

4 Layers of sediment can pile up for millions of years. The sediment at the bottom of the pile is squashed. The grains become cemented together. A new rock called **sedimentary rock** is formed.

Sandstone was made in this way from small grains of sand. **Mudstone**, **shale**, and **clay** were made from even smaller grains. **Limestone** and **chalk** are also sedimentary rocks, but the sediment which made them was rather unusual. They were made from the shells and skeletons of tiny prehistoric sea animals, which died, fell to the sea bed and were buried and squashed.

Sediment at the mouth of a river

1 What is meant by the erosion of a mountain? What causes it? ▲
2 Describe how bits of rock from a mountain can get into the sea. ▲
3 What is sediment made up of? How is sedimentary rock made? ▲
4 Limestone is a sedimentary rock. Suggest how it was formed.
5 Why is river sand rougher than beach sand? ▲
6 Two sedimentary rocks are described on page 46. Read about them, then write down things which sedimentary rocks have in common.
7 A large lorry can carry 10 tonnes. How many lorries could be filled by the sediment carried down by the Mississippi in one day?

Did you know?

- The Mississippi river carries over 500 million tonnes of sediment to the sea each year.
- Grains and pebbles are worn smooth as the water carries them along. That's why beach sand, which is always on the move, is smoother than river sand.

49

Many statues are made of marble. It's hard, attractive rock made of crystals which sparkle in the light. You wouldn't use limestone for a statue. It's made of grains which are easily scratched off and is not at all shiny.

Marble and limestone seem to be very different, but they both 'fizz' when dropped into acid. That's because they are made of the same mineral – calcite. In fact, marble started off as limestone.

Marble is a **metamorphic rock**, a rock which has been changed from its original form.

The diagram shows one way in which marble could have been made from limestone. Hot magma has pushed its way upwards into beds of sedimentary rock. As the molten rock cooled and solidified, lots of heat was given out. This heat baked the rock next to it. The calcite grains were changed into calcite crystals, changing the limestone into marble. Grainy sandstone changed into crystalline **quartzite** in the same way.

From this... a piece of limestone

to this ? a marble statue

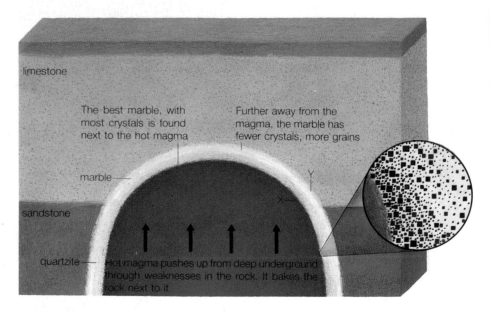
limestone

The best marble, with most crystals is found next to the hot magma

Further away from the magma, the marble has fewer crystals, more grains

marble

sandstone

Y

X

quartzite   Hot magma pushes up from deep underground through weaknesses in the rock. It bakes the rock next to it

High pressures can also change rocks. Some **slate**, for example, was made when mudstone was squashed deep underground. Under pressure, the tiny mudstone grains were changed to flat crystals. These crystals grew along layers. That's why slate can be split easily.

The biggest regions of metamorphic rock were formed by the action of **high temperature** and **pressure**. In these regions, Earth movements crumpled and folded the crust into mountains. During this mountain building huge underground temperatures and pressures were produced and whole areas of rock were changed.

1   What is a metamorphic rock? ▲
2   Write down three kinds of metamorphic rock mentioned here. ▲
3   Describe one way in which: a) marble   b) slate could have been made. ▲
4   Why do limestone and marble both fizz in acid? ▲
5   The marble at X is mostly made of crystals. The marble at Y has crystals and grains. Explain why there is a difference.
6   **Try to find out:** why metamorphic rocks don't normally contain fossils.

Did you know?

● Heat and pressure, working together, can bend and twist rocks.
● At a depth of 10 km, rocks are affected by a pressure of 20 000 kg on each square centimetre of rock.

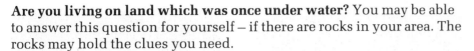
Are you living on land which was once under water? You may be able to answer this question for yourself – if there are rocks in your area. The rocks may hold the clues you need.

The rocks may clearly show grains of sand or gravel or pebbles. In that case you can be fairly certain that you are standing on land which was formed under water, from sediment.

The rocks may be made up of layers. This is also good evidence that they were formed under water. When sediment settles out on the bed of a lake or sea, it forms layers. When the sediment is squashed into rock, the rock is also formed in layers.

The rock may contain fossils. Fossils are traces of prehistoric life. Most were formed from the remains of animals and plants which were buried in the sediment. Sometimes parts of the animal or plant were preserved. More often, the parts were replaced by minerals which filled in exactly the same shape. The type of fossil found in an area gives information about the area's history. Your local rocks, for example, may contain fossils of **corals**, animals which live in shallow seas. If the rocks *do* have **coral** fossils, you are probably living on a prehistoric sea bed!

A rock made of sand, gravel and pebbles

Rock in layers. The Grand Canyon, USA

A coral fossil

A dinosaur footprint

Studies by geologists suggest that much of Britain has been under the sea several times. Great earth movements brought the sedimentary rocks to the surface, making much of Britain's 'dry land'.

1 Write down three clues which might tell you that a rock was formed under water. ▲
2 Why is sedimentary rock often made up of layers? ▲
3 What is a fossil? How were fossils formed? ▲
4 A quarry has lots of coral fossils. What does this tell you about the rock in the quarry? Explain your answer.
5 Try to explain: a) why sea shell fossils are found on top of Mount Everest   b) why igneous rocks don't contain fossils.
6 **Try to find out:** if any fossils have been found near your town.

Did you know?

- The grains in some sedimentary rock are too small to see – even with a microscope.
- There are fossil sea shells near the top of Mount Everest.
- Some sandstones show fossilised tracks of animals which crawled over them.

In the old parts of most towns, you can see houses made of stone. For thousands of years, man has used rocks for building.

Slate has probably been used in more British buildings than any other rock. It's a good building material. It is hard and strong and isn't affected by weather or chemical pollution. It is particularly useful because it can be split into flat sheets. These have been used for making fences and pavements and, most of all, for making roofs for houses.

Roofing slates were carried all over Britain from quarries in North Wales, Cornwall, and the Lake District. But the stone used for building house walls wasn't carried nearly so far. Builders normally used stone from the local quarry. That's why buildings in different parts of the country are made of different stone.

Splitting slate

Granite church, Aberdeen

Sandstone house, Northamptonshire

Limestone cottage, Gloucestershire

Granite is expensive to cut and shape because it is so hard. On the other hand, it is not affected by weather or pollution. It keeps its shiny appearance.

Since **sandstone** is much softer than granite, it is cheaper to work. But soft sandstone does not stand up well to wind and rain. Where there is pollution, it becomes dirty.

Limestone is another soft, easily worked rock. Acid in rain can keep limestone surfaces clean by dissolving a little of it. But rain with lots of acid eats away a limestone building.

Did you know?

● Some limestones are so soft that they can be sawn by hand.
● Some slabs of slate are flat enough for making billiard tables.

1 a) Why is slate a good building material? ▲
   b) Why is slate useful for roofing houses? ▲
2 Aberdeen has many granite buildings. In Yorkshire, limestone is the main building stone. Why is there a difference?
3 Explain why rain can improve or spoil a limestone building. ▲
4 Which phrases describe: a) limestone b) sandstone
   c) granite?
   easy to shape   keeps its appearance   not affected by weather
5 Suggest: a) which of the two houses in the picture was more expensive to build b) why the lower house has lasted better than the cottage.
6 **Try to find out:** which stone is produced by your nearest quarry.

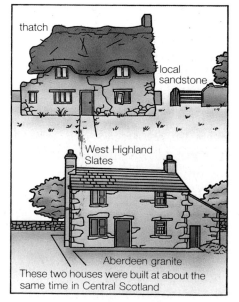
thatch
local sandstone
West Highland Slates
Aberdeen granite
These two houses were built at about the same time in Central Scotland

Going further

The Pyramids, in Egypt, were built from natural stone. They have lasted well! But little stone is used for building nowadays. It costs far too much to shape and transport it.

One rock, however, does have a large part to play in modern building. That rock is **clay**.

Clay

Clay is a sedimentary rock made of very fine grains. When dry, it is very brittle. When wet, it is like putty or Plasticine. Some people *do* live in huts covered with clay, but really it is quite unsuitable for building.

Firing clay, however, changes that. Fired clay, can be used to make bricks, roof tiles, drainpipes and chimney pots. The wet clay is cut or moulded into shape, then allowed to dry. Finally it is fired – baked in huge ovens called **kilns**. This drives off the last of the water leaving a hard strong solid.

Cement

Clay is also used to make another important building material – **cement**. The cement is made by roasting clay and limestone together, then grinding down the mixture to a powder.

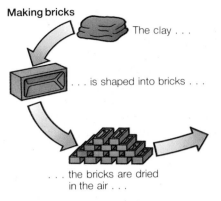
Making bricks
The clay . . .
. . . is shaped into bricks . . .
. . . the bricks are dried in the air . . .

then fired in a kiln

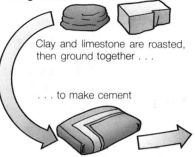
Making cement
Clay and limestone are roasted, then ground together . . .
. . . to make cement

Mixing the cement with sand, gravel, and water makes concrete

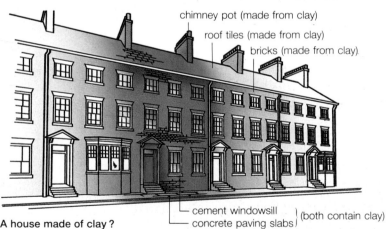
chimney pot (made from clay)
roof tiles (made from clay)
bricks (made from clay)
cement windowsill
concrete paving slabs
(both contain clay)
A house made of clay ?

When wet cement dries, it sets to a hard solid. That's what makes it so useful. Bricklayers use **mortar**, a mixture of cement, water, and sand, to join bricks together. **Concrete**, which contains cement, sand, water and gravel, is used in most modern buildings.

1   What is clay? How was it formed in nature? ▲
2   a) Why is natural clay unsuitable for building? ▲
    b) Why is fired clay a much better building material?
3   What is meant by: a) firing clay   b) a kiln   c) mortar? ▲
4   Suggest: a) how a roof tile is made from clay   b) why clay roof tiles are unlikely to become as scarce as slate ones.
5   Why should bags of cement be stored in dry conditions?
6   **Try to find out:** something about china clay.

Did you know?

- Bricks have been made for more than 5000 years.
- Clay which is suitable for making bricks is found quite commonly.
- Steel rods are often put into concrete beams to strengthen them.

In the Yorkshire Dales, and in Somerset, you can find some rather unusual countryside. You can find:

- rivers which suddenly disappear underground
- deep gorges which cut through pleasant grassland
- caves which stretch underground for several kilometres.

This scenery is typical of limestone country. Limestone rocks are reactive rocks, and that's because they are made of **carbonate**.

Carbonates – and acids

Carbonates are chemical compounds which contain carbon and oxygen joined together. Limestone is made of calcium carbonate which contains calcium, carbon, and oxygen.

All carbonates are reactive. For example, they **react easily with acid**.

When dilute hydrochloric acid is added to a carbonate, the carbonate reacts and carbon dioxide gas is made. (The 'fizz test' on page 46.)

Carbonates also react with rain, which is very slightly acid, even where there is no pollution. This reaction does not make carbon dioxide, but it does very slowly change the carbonate into a soluble compound. This is the reaction which keeps limestone buildings clean (page 52). It has also, over millions of years, produced the spectacular limestone scenery. As the rain and river water have slowly worn away the limestone rock, they have produced gorges, caves, and potholes.

Heating carbonates

Carbon dioxide can be produced from limestone in another way – by roasting it. Most carbonates give off carbon dioxide on heating.

When limestone is roasted, **quicklime** is also produced. This word equation shows what happens:

calcium carbonate → calcium oxide + carbon dioxide
   (limestone)   (quicklime)

Some of the lime used in farming is produced in this way by heating limestone to 1200°C in a **lime kiln**. This reaction also takes place as a first step in the making of cement.

Limestone country in Yorkshire

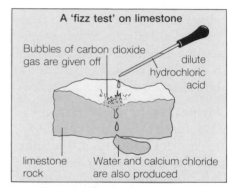

A 'fizz test' on limestone

Bubbles of carbon dioxide gas are given off

dilute hydrochloric acid

limestone rock

Water and calcium chloride are also produced

A marble statue, badly affected by air pollution

1   What is: a) a carbonate   b) the chemical name for limestone? ▲
2   How have potholes and caves been made in limestone country? ▲
3   What is made when limestone is roasted? ▲
4   How could you show that carbon dioxide is given off in the 'fizz test' shown above? Name the other chemicals produced, and the chemicals which react, then write a word equation.
   _____ + _____ → _____ + _____ + calcium chloride
5   Is it possible that acid rain has damaged the statue in the photograph? Explain your answer.
6   **Try to find out:** what stalactites and stalagmites are.

Did you know?

- Chalk, limestone, and marble are all made of calcium carbonate.
- About 50 million tonnes of limestone are used in Britain each year, mostly for making cement.

12.3 The Rock Cycle goes round and round For the enthusiast 2

When you look at a mountain it is very easy to think that it will never change. It looks as though it has been there for ever and will always be there. However rocks, like everything else, will eventually be worn down by processes of weathering and erosion.

Igneous rocks are broken down into fine particles which are carried away by rivers to be deposited and eventually form sedimentary rocks. These rocks may be slowly buried deep in the Earth's crust where they are affected by very high temperatures and great pressure (metamorphism). Metamorphic rocks are gradually forced to the surface by movements within the Earth's crust. Some will be pushed even further down in the Earth until they melt completely. They form molten rock in the mantle. Some of this molten rock will rise to the surface of the Earth and spill out of volcanoes.

As rocks emerge from deep within the crust they are once again subjected to the effects of weathering and erosion.

And so the rock cycle goes on and on …

Will this mountain look like this for ever?

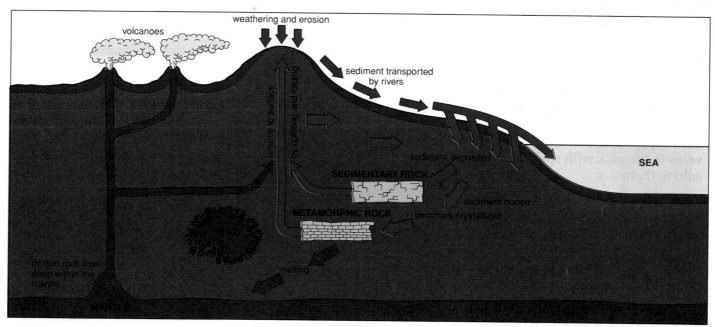

The rock cycle

1   Write down two ways in which mountains are worn down into tiny particles. ▲
2   Why is water important in the 'recycling' of rocks? ▲
3   Describe how metamorphic rocks get to the surface of the Earth's crust. ▲
4   What happens to rocks when they reach the mantle? ▲
5   Explain why the rock cycle is called a cycle. ▲
6   Which kind of rocks will take the: a) shortest   b) longest time to be worn down into sediment?
7   **Try to find out:** how long it takes for molten rock from a volcano to cool into solid rock.

Did you know?

- Rocks in North Wales came from volcanoes that erupted 500 million years ago.
- If the Earth was filmed using time-lapse photography it would be 100 million years before we would notice any change in the shape of mountains.

12.4   How long will metals last?

Supplies of metal ores will not last forever. The table gives information about how long reserves of some metals are likely to last. As you can see, the reserves of some metals will last longer than others. But, it is obvious that if we are not very careful all natural reserves of the different metals will eventually run out. We must try to conserve valuable resources for as long as we can.

We can do this by looking for alternative materials such as plastics or by **recycling** waste materials.

Recycling means using waste materials to make new ones.

All metals can be recycled. The scrap industry recycles large amounts of metal every year. Half of the aluminium used today is recycled. This process uses only 5% of the energy used to extract aluminium from its compounds by electrolysis. Fifty per cent (50%) of copper is recycled from old pipes and wire. Producing copper from scrap costs only 3% as much as mining it from the Earth.

So, recycling makes sense!

What can you do to help?

It is easier to recycle big pieces of metal like an old car body, than small ones such as drink cans (that we throw away almost every day). Drink cans are made from either pure aluminium or tin plate (steel with a thin layer of tin). Both of these metals are running out fast so recycling is very worthwhile. The scrap metal has to be collected, transported, and separated carefully. All this can be very expensive, so expensive in fact that recycling may not be worthwhile.

We can all help by separating our drinks cans and taking them to a recycling centre. It is easy to tell aluminium from tin cans by using a magnet - tin is magnetic, aluminium is not.

Metals are not the only materials that can be recycled.

Glass, paper, oil, plastics, and even old clothes can all be used over again. More of the valuable Earth's resources are saved and it cuts down the amount of rubbish to be disposed of.

aluminium	30 yrs
copper	74 yrs
gold	27 yrs
iron	302 yrs
lead	49 yrs
tin	40 yrs
zinc	38 yrs

How long reserves of different metals will last

Recycling makes sense

Did you know?

- About one tonne of rubbish is thrown away by the average family each year.
- Of the 600 000 000 glass bottles and jars thrown away each year only 10% are recycled.
- Almost all of the gold used today is recycled. People don't throw their old jewellery away.

1   What does recycling mean? ▲
2   Give some examples of material which can be recycled. ▲
3   Why is the recycling of metals important? ▲
4   Explain why a magnet can help you to tell whether a drinks can is made from aluminium or tin plate. ▲
5   Plastics are made from oil. Most plastic is not biodegradable and will not 'rot' away. Suggest why the recycling of plastic is good for our environment.
6   **Try to find out:** where your nearest recycling centre is. What materials are recycled there?

12.5   Fuels from the Earth

Coal   This is how scientists think that coal was formed: Millions of years ago large areas of Britain were swamps. Thick forests grew there. When dead trees fell into the swamps, they were buried by mud.

The mud kept the air out and so the trees did not rot away. As time passed, however, the woody material did change. As the mud piled up, the trees were squashed together. Heat from inside the Earth also affected them. The trees were changed into coal.

The dead trees were buried in the mud

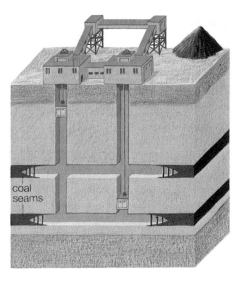

coal seams

Now much of this coal lies deep underground, covered by large amounts of sediment. To mine these underground coal seams, deep shafts have to be sunk. Some coal seams, however, have been pushed towards the surface. This coal can be mined more easily once the surface rock has been bulldozed away.

Oil   This is how scientists think that oil was formed: Long ago, the North Sea was warmer and shallower than it is now. Countless millions of tiny animals and plants lived in it. When these organisms died, their bodies fell to the sea bed. There they were trapped in sand and mud which formed as sediment.

Pressure changed the sand and mud to sandstone and mudstone. Pressure, and heat, changed the organisms to oil and gas.

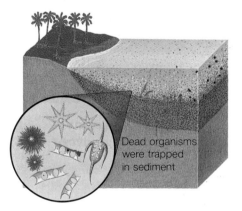

Dead organisms were trapped in sediment

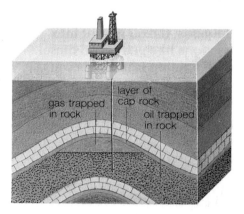

gas trapped in rock | layer of cap rock | oil trapped in rock

The underground pressure forced the oil (and gas) upwards, through the rock. Some oil escaped to the surface, but in many places it was trapped under layers of **caprock** which did not let it through. These areas are now oilfields. The oil is trapped as tiny drops between the rock grains. When an oil rig drills through the caprock, the oil is released.

1   a) What were coal and oil formed from? ▲
    b) How were these fuels formed? ▲
2   Why is coal a fossil fuel? Name another fossil fuel. ▲
3   Oil is not found in underground pools. How is it found? ▲
4   Sometimes, oil can rise to the surface without pumping. Suggest why this happens.
5   The energy stored in coal came from the Sun. Explain this.
6   **Try to find out:** a) what peat is, and how it was formed
    b) which countries have large coalfields or oilfields.

Did you know?

- Coal is called a **fossil fuel** because it was made from things which lived on the Earth long ago.
- The oil which comes out of the Earth is hot. It can be as hot as 80 °C.

12.7 Salt from the sea

The Dead Sea lies between Israel and Jordan, completely surrounded by land. It's the most salty sea in the world – so salty that you can float in it sitting up! It's far too salty for plants and animals to live in.

Where does all the salt come from? Like all sea salt, it is carried from the land to the sea by rivers. Rain falling on high ground to the north of the Dead Sea dissolves tiny amounts of chemicals from the rocks and soil there. These soluble chemicals (called **salts**) are washed into the rivers which run into the Sea. The water becomes more and more salty. The very hot weather evaporates the sea water as quickly as the river water flows in. The salt is left behind.

All seas become salty in this way. Normal sea water contains around 25 g salts dissolved in each kilogramme of sea water. 1 kg of Dead Sea water contains over 200 g!

The dead sea is so salty that you can sit up in it

Millions of tonnes of salt are produced each year by evaporating sea water. In many places, heat energy from the Sun evaporates the water. In the lab, you can use a bunsen burner.

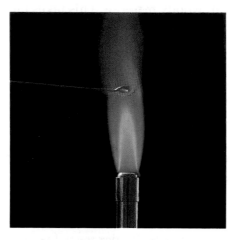

If you put a little sea salt into a flame, the flame turns orange. This shows that sea salt contains **sodium**.

If you pass an electric current through sea water, **chlorine gas** is made. Sea salt contains chlorine. The main compound in sea salt is **sodium chloride**.

What is the salt used for?

Salt is used for cooking, and on roads in winter. But most salt is used by industry to make chemicals like **chlorine** (used in swimming pools) and **sodium** (its vapour is used in street lights).

1 How did the salt get into the Dead Sea? ▲
2 Suggest: a) how the Dead Sea got its name   b) one reason why the Dead Sea water is so salty. ▲
3 a) How could you find the mass of salt in 100 g of sea water? What mass would you expect to find?
   b) How could you show that sea salt contains sodium chloride? ▲
4 Some places have deep underground salt beds. Water is pumped down to the salt and back again. Suggest a reason for doing this.
5 **Try to find out:** as much as you can about the Great Salt Lake in Utah, U.S.A.

Did you know?

- The Dead Sea is the world's lowest sea. Its surface is 400 m below the Mediterranean.
- Roman soldiers used to be paid in salt. That's where the word 'salary' comes from.

Treasure store?

The chemicals found in sea water include compounds of around 75 different elements. Among these elements are:

Element	silicon	aluminium	magnesium	gold	calcium
% in Earth's crust	27.7%	8.1%	2.4%	0.000000005%	3.22%
% in sea water	0.0003%	0.00001%	0.12%	0.0000000004%	0.04%

Element	sodium	potassium	chlorine	iodine	bromine
% in Earth's crust	2.8%	2.5%	0.03%	0.00003%	0.0002%
% in sea water	1.1%	0.03%	1.9%	0.00005%	0.006%

There is plenty of sea water, and so you might expect the sea to provide a rich harvest of elements. In fact, only a few elements like sodium, magnesium, bromine, and chlorine are extracted (separated) from sea water. To obtain reasonably large amounts of the other elements, huge volumes of water would have to be treated. This makes the extraction far too expensive to be worthwhile.

Iodine is also extracted – with the help of nature! Seaweeds build up iodine compounds in their tissues. The seaweed is harvested to obtain the iodine.

Dump?

Some other substances in the sea are much less welcome. Sea water contains poisonous and radioactive chemicals, as well as untreated sewage.

Many people think that they can safely dump waste materials in the sea. They think that the oceans are so large that the harmful material will be carried away from land. But this thinking can be badly wrong. A few examples show why:

- Dumped sewage has turned many Mediterranean beaches into health hazards.
- Dumped radioactive chemicals have been washed up in Cumbria.
- Pollution has killed off iodine-producing seaweed in California.
- Dumped toxic metals are absorbed by fish. This caused a disaster in Japan in 1951. A Japanese company solved its waste mercury problem by dumping the waste offshore. But fish absorbed the mercury in large quantities. People who ate those fish suffered horrific illness. Some even died.

1 a) How many elements have been found in sea water? ▲
   b) Why are so few metals extracted from sea water? ▲
2 Chemical treatment of sea water can produce magnesium chloride. How can magnesium be obtained from this compound? (See p. 57.)
3 Make a list of things which can pollute the sea. ▲
4 Using the table at the top of the page write down: a) the five most plentiful elements in the sea   b) three elements which are more plentiful in the sea than on land.
5 Work out: a) the masses of chlorine and gold in 100 g of sea water
   b) the masses of sodium and magnesium in 1 kg of sea water.
6 **Try to find out:** how oil pollution is dealt with.

From this...
iodine – containing seaweed

...to this
a quartz-iodine car headlamp bulb

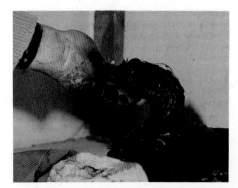

Each year, thousands of birds are killed by oil dumped in the sea

Forces and movement

A pile of slabs

+ a mighty force

= a pile of rubble...

It's hard to believe that a man can smash paving stones with his bare hands, but this karate expert, Colie MacLeod, can do just that. He once struck a pile of slabs with such a great force that he smashed 26 of them.

There's nothing unusual about a human body producing forces. Your body produces forces all the time, allowing you to walk, to lift things, to jump, to climb, and much more. What was unusual about Colie's feat was the size of the force he used. Many people smash slabs, but not with bare hands. The normal thing to do is to use a machine which allows the work to be done more easily.

You won't learn anything about paving slabs in this section, but you will learn a lot about forces – what they are, what they do, and how your body produces them. And you will learn something about simple machines, including one which could help with slab smashing. See if you can recognise which one it is!

A force is a push or a pull.

You can use a **pushing** force to move a trolley.

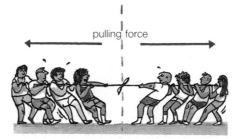

In a tug of war, you use a **pulling** force, if both teams pull with the same force there is no movement.

You normally use a **push** and a **pull** to turn the handlebars of your bicycle.

Forces can make things change shape, change direction, change speed.

The springs have changed shape

A strong man can produce a big enough force to stretch a chest expander or squash a can out of shape.

The ball has moved away from the racket in a different direction

The force from a tennis player's racquet can change the direction in which the ball is travelling.

The force produced by the brakes has slowed down the car's wheels

The force produced by a car engine makes the car begin to move, then speed up. A force from the brakes can slow it down, then stop it.

Two forces can help each other. It's difficult to move a car by pushing it yourself. It's easier if a friend helps you to push in the same direction.

Two forces can work against each other. That's what makes a tug of war so tiring. Your team pulls one way. The other team pulls in the opposite direction.

1 What is a force? What can forces do? ▲
2 What kind of force do you use to: a) move a trolley b) win a tug-of-war c) squash a can d) climb a rope?
3 Make a list of machines which have been designed to produce pushing (or pulling) forces.
4 When a footballer heads a ball, his head puts a force on it. This force affects the ball in three ways. What are they?
5 If a car crashes into a wall, the wall puts a force on the car. The car puts an equal force on the wall. What does each force do?
6 **Try to find out:** what is meant by balanced forces.

Did you know?

● An American strongman called Paul Anderson once produced a force big enough to lift three Rover 100 cars.
● The longest tug of war ever recorded lasted 2 hours 41 minutes. It took place in India in 1889, between two teams of British soldiers. The record is unlikely to be broken. The soldiers were able to lie on the ground, and this is not now allowed.

Here are some of the different forces which act on or around you.

up thrust

A **force of friction** is produced whenever two surfaces rub on each other. The force of friction slows down things which are moving.

Everything on Earth, and on other planets, is affected by the force of **gravity**. Gravity pulls everything downwards.

A piece of wood floats on water. It's kept up by a force called **upthrust**. The force is produced by water pushing on the wood from underneath.

The steels pins are attracted to the magnet by a **magnetic force**. The magnet can attract anything which contains iron, steel, cobalt, or nickel.

A raindrop keeps its shape when it lands on a waterproof surface. That's because of **surface tension**. The water molecules attract each other strongly and form a kind of skin on the water's surface.

You can pick up paper with your plastic pen – if you rub it on a duster first. Rubbing gives the plastic an electric charge. An **electric force** pulls the paper to the pen.

1 Which of the six forces: a) keeps a piece of wood floating b) keeps a raindrop in shape c) slows down things which are moving? ▲
2 Which of the six forces: a) picks up dust on a record b) keeps the Moon in orbit round the Earth c) helps you when swimming?
3 a) Which substances are attracted to a magnet? ▲
b) Why can a magnet pick up a tin can? (page 20 will help)
4 What does the force of gravity do? ▲
5 Explain carefully how gravity affects a firework rocket: a) when it is on the way up b) when it is coming down.
6 **Try to find out:** why upthrust is important to the whale, and why the whale gets into serious trouble when it is beached.

Did you know?

- You can float a pin on water – if you lay it carefully on the surface. Surface tension keeps it up.
- Gases and liquids can both supply upthrust. Upthrust is the force which makes a hydrogen balloon rise when you release it.

13.1 How do they work?

On this page you will see four devices. Each uses one or more of the forces you met on page 70. How do the devices work?

1 You would find this hanging from a crane in a scrap yard. With the electricity switched on, it can pick up a motor car, but not a copper boiler.

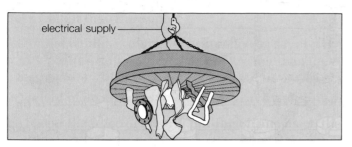

a) Which force does it use? Give a reason for your answer.
b) How does it work? (*Clue*: the electricity is switched on and off.)

2 This can be used to separate minerals from a finely ground ore.

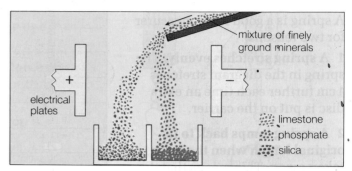

a) Which force does it use to separate the minerals?
b) How does it work? (*Clue*: the minerals become electrically charged during grinding.)

3 This is a piece of history. It was once used for carrying slate downhill. Trucks A and B ran side by side on two railways. Truck A carried a water tank and truck B the slate.
a) Which force did the machine use?
b) How did it work? (*Clue*: at the end of one run, the water tank was emptied. At the end of the next, it was filled.)

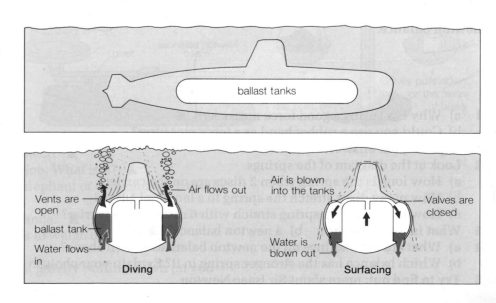

4 A submarine has huge tanks, ballast tanks, running along its length. They have holes at the bottom and vents at the top. Along with bottles of air and air fans, these tanks control the submarine's up and down movement.
a) Which force pulls the submarine down?
b) Which force pushes it up?
c) How would you make the submarine rise? How would you make it dive? (*Clue*: look to see what is happening in the tanks.)
d) What can you say about the sizes of the forces when it is rising, and when it is diving?

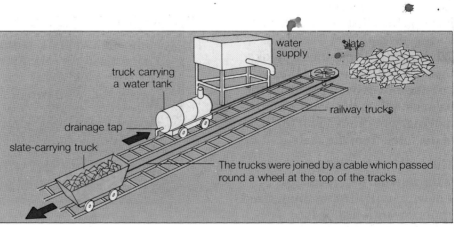

A see-saw and a beam balance have much in common. If you can understand how one works, you can understand the other.

A see-saw

Two people have to sit on a see-saw before it will balance. They don't have to be the same weight – a fat person and a thin person can make the see-saw balance. But it does matter where they sit.

Look at the two see-saws on the right. Whether each see-saw balances or not depends on:

- **the size of the force on each side.**
- **the distance between each force and the pivot.**

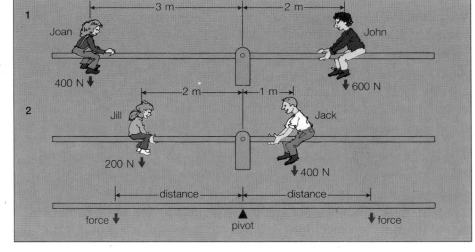

The see-saw balances when . . .

$$\text{force} \times \text{distance} = \text{force} \times \text{distance}$$
$$\text{(on the left of the pivot)} \quad \text{(on the right of the pivot)}$$

For see-saw **1** $400\,\text{N} \times 3\,\text{m} = 600\,\text{N} \times 2\,\text{m}$

A beam balance

A chemical balance is a bit like a see-saw. It is made of a swinging beam balanced on a pivot. On one side of the pivot, the beam carries a weighing pan. On the other side, it carries a metal rider which can be moved backwards and forwards. This side of the beam has a scale marked on it.

To weigh an object on the balance, you have to:

- put the object on the weighing pan
- slide the rider until the beam balances
- read the weight off the scale.

A steelyard works in the same way. So do many of the weighing machines in doctors' surgeries.

The stone is heavier than the metal rider, but it is still possible to make the beam balance by sliding the rider to the right

1   What do a see-saw and a chemical balance have in common? ▲
2   Whether or not a see-saw balances depends on two things.
    a) What are they?   b) When will the see-saw balance? ▲
3   How would you find the weight of your pencil using a chemical balance?
4   Look at the see-saw pictures above.
    a) Fred takes Joan's place on see-saw 1. He weighs 300 N. Where should he sit to balance? (Draw a picture – it will help.)
    b) Jack moves backwards on see-saw 2 until he is 1.5 m from the pivot. Where should Jill move to keep balanced?

Did you know?

- Two Americans once see-sawed non-stop for 44 days.
- Chemical balances have pivots made of special toughened steel, sharpened to a knife edge. This cuts down friction and makes the balance more accurate.

13.2 Kilograms and newtons

If you think about it, something seems to be wrong. On page 72 you learned that weight is measured in newtons. And yet weighing machines in shops, markets, and even in science labs, have scales marked in grams and kilograms, not in newtons.

In fact, something is wrong. Grams and kilograms are units of mass, not weight – mass and weight are different! If you can use your imagination, and your brain, you should be able to work out

- something about mass
- something about the difference between mass and weight as you read through this page.

You'll have to use your imagination to picture two tins of beans (HB57 and HB58) on very long voyages. The tins are identical. Each contains 1 kg of beans. Each sat on the same shelf of a supermarket in Edinburgh until ... HB57 was bought by an explorer going to the North Pole ... HB58 was bought by an astronaut going to the Moon. The diagram shows the mass and weight of the beans in each can at different stages of the journeys.

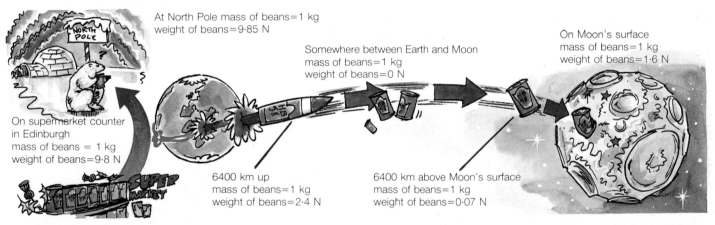

At North Pole mass of beans = 1 kg
weight of beans = 9·85 N

Somewhere between Earth and Moon
mass of beans = 1 kg
weight of beans = 0 N

On Moon's surface
mass of beans = 1 kg
weight of beans = 1·6 N

On supermarket counter in Edinburgh
mass of beans = 1 kg
weight of beans = 9·8 N

6400 km up
mass of beans = 1 kg
weight of beans = 2·4 N

6400 km above Moon's surface
mass of beans = 1 kg
weight of beans = 0·07 N

1 a) Why is weight measured in newtons?
b) What do you notice about the weights of the beans at different stages of the journeys?
c) What is meant by saying 'on the Moon, the weight of the beans in the HB58 can is 1.6 N?
d) Suggest why the HB58 beans weigh less on the Moon than on Earth, and why they weigh zero at one point in space.

2 1 kg is a measure of the quantity of beans in each can.
a) What do you notice about the mass of the beans on the journeys?
b) HB57 was left behind at the North Pole. The mass of beans in the can was zero. What had happened?

3 To make calculations easier, the weight of a 1 kg object is taken to be 10 N. Using this, look back to page 72 and work out:
a) the mass of the apple and the bag of potatoes
b) their weights on the Moon.

4 **Try to find out:**   a) why zero gravity causes problems for astronauts   b) what mass is.

Did you know?

- On Earth, gravity pulls down more strongly at the Poles than at the Equator.
- The force of gravity is different on different planets. Jupiter's gravity is 2.3 times greater than Earth's. Earth's is 3 times greater than Mars'.
- HB58 (and the rocket which carried it!) would have have to reach a speed of 11 000 metres per second to escape from Earth's gravity.

13.3 The force of friction

There is no such thing as a completely smooth surface. There are some highly polished surfaces which seem to be perfectly smooth. But if you look at these surfaces with a microscope, you can see that even they have rough edges. The rubbing of these rough edges causes the force called **friction**.

Friction is the force produced when two surfaces rub on each other. Friction tries to stop the surfaces from sliding over one another. The force of friction is small for fairly smooth surfaces like glass and ice. It is much greater for rough surfaces like sand paper or concrete.

Friction can be useful. You couldn't walk without friction. Friction prevents the soles of your shoes from slipping over the ground.

When you write with a pencil, friction rubs millions of carbon atoms off the end of the pencil . That's what leaves the black mark on the paper.

It's often useful to make friction as large as possible. That's why climbers use rubber soled boots. They produce more friction and give a better grip.

Friction can cause problems. Then it has to be made smaller. When the moving parts of machines rub on each other, they are worn away by friction. The machine is slowed down too. That's why machines have to be **lubricated** with oil or grease. Lubricating cuts down friction. A well oiled machine runs more smoothly, and lasts longer.

Wheels are often used to cut down friction.

Feel how smooth the page is–
this is what it looks like under a microscope

Using friction—for climbing

Using friction—for writing

Cutting down friction—to make the wheels run smoothly

Cutting down friction—for faster skiing

1   What is friction, what causes it and what does it do? ▲
2   Which surfaces produce: a) most friction   b) least friction? ▲
3   What is meant by lubricating? Why is it important? ▲
4   Why is it not much fun to ride a badly lubricated bicycle?
5   a) Why do climbers wear rubber soled boots? ▲
    b) Why do ballroom dancers wear leather soled shoes?
6   Suggest why: a) wheels are so useful   b) a pencil will write on paper but not on glass   c) climbing a greasy pole is so difficult.
7   **Try to find out:** how and why friction is cut down
    a) in a hovercraft   b) on a dance floor   c) in a car engine.

Did you know?

- Skiers wax their skis to cut down friction.
- Teflon, which is used to coat non-stick frying pans, produces less friction than any other solid.

13.3 Friction on the roads

Driving a motor car would be impossible without friction.

Tyres grip the road by friction and that's what allows a driver to control his car. There has to be a large amount of friction between the tyres and the road so that the tyres can grip the road well. Then:

- when the engine turns the wheels, the car can go forward
- when the driver turns the steering wheel, the car can turn a corner
- when the driver puts the brakes on, the car can stop.

Brakes work by friction too. There is a set of brake pads on each of the car's wheels. (They are a bit like the brake blocks on your bicycle.) When the driver presses on the brake pedal, these brake pads press on metal discs fitted behind the wheels. This produces friction between the pads and the discs, making the wheels slow down.

If the friction between the tyres and the road is reduced, driving becomes dangerous. That's why drivers have to take great care when driving on icy roads. Driving on wet roads also needs extra care. The water acts as a lubricant between the tyres and the road and makes braking more difficult. The diagram below gives you some idea of how braking distances are affected by wet conditions.

Tyres grip the road by friction. But some do this better than others

If the driver of a car travelling at 80 km per hour (50 mph) wants to stop his car:

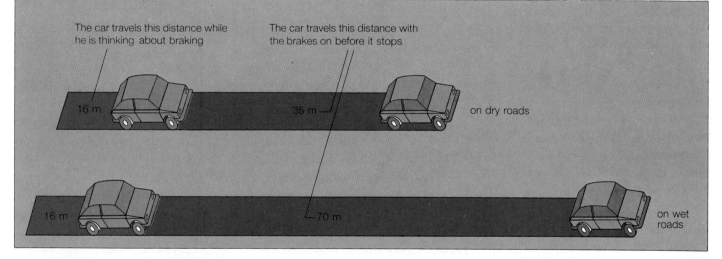
The car travels this distance while he is thinking about braking — 16 m. The car travels this distance with the brakes on before it stops — 35 m on dry roads. 16 m — 70 m on wet roads.

1 Why it is important that there are large amounts of friction:
a) between car tyres and the road b) between brake pads and the wheel? ▲
2 a) Car wheels often spin on icy roads. Explain why.
b) Why are sand and grit spread on roads in winter?
3 Why is braking more difficult on wet roads than on dry? ▲
4 A man is driving along a motorway at 80 km per hour. What distances should he leave between his car and the car in front:
a) in dry conditions b) in wet conditions?
5 The photo above shows a normal tyre, an illegal tyre, and a winter tyre. Explain which is which.
6 **Try to find out:** about tyres used by racing drivers.

Did you know?

- It is illegal to drive a car with a tyre tread less than 1.6 mm deep.
- The pattern of grooves on a tyre is designed to let water escape from underneath it.

13.3 So fast, but no faster

To make a bicycle go, you have to supply a force. You push on the pedals and this force turns the wheels. With your first push, the bicycle moves off. As you keep pedalling, the bicycle speeds up. This fits in with what you learned on page 69:

A force can make something start to move, then speed up.

But the bicycle doesn't keep going faster and faster as you keep pedalling. Instead, it reaches a steady top speed. Then you have to continue pedalling just to keep it going at that speed. You are using a force without changing the bicycle's speed.

Why does the bicycle not go faster and faster?

The quick answer to that question is 'friction'. When you are cycling, two types of friction hold you back. The first is caused by the rubbing of moving parts, like the wheel and the axle. The second is caused by the air. To cycle, you have to push your way through the air. It sets up the force of friction called **air resistance**. These friction forces get bigger the faster you go.

At low speed, the friction is small, much smaller than the force pushing the bicycle forward. And so most of the force which you put on the pedal goes towards speeding up the bicycle.

But, as the bicycle speeds up, friction increases. Eventually it gets so large that it is equal to the force pushing the bicycle forward. All your force is used to overcome friction. The two forces are **balanced** (equal and opposite). The bicycle keeps going at a steady speed.

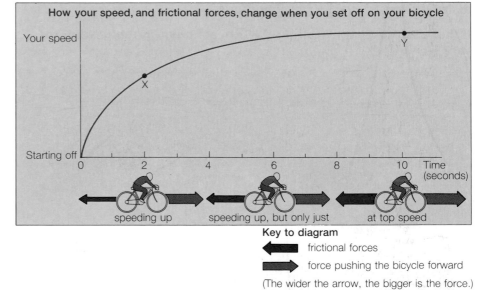

How your speed, and frictional forces, change when you set off on your bicycle

Key to diagram
frictional forces
force pushing the bicycle forward
(The wider the arrow, the bigger is the force.)

And where is there no friction?

A space rocket can keep flying at a steady speed with its engines off because there is no friction in space to slow it down.

1   When you cycle, which force: a) moves you forward
    b) holds you back? ▲
2   Why does your bicycle: a) speed up when you start pedalling
    b) slow down when you stop pedalling? ▲
3   Why does the bicycle not go faster and faster as you pedal? ▲
4   Why can a space rocket fly at a steady speed with its engines off? ▲
5   Why is the cyclist (in the diagram): a) speeding up at X
    b) going at a steady speed at Y? (*Clue:* think about forces.)
6   How long does the cyclist take to reach top speed?
7   **Try for find out:** how a space rocket in space can be slowed down.

Did you know?

- When a bicycle's speed doubles, air resistance increases four times.
- Space rockets fly for most of the time without burning their engines. The rockets speeds can very gradually change, however. They are affected by the pull of gravity from the nearest planet.

13.3 Free fall

'Ready, Joan?' The instructor points to the aircraft's door.

I sit down on the sill, legs dangling outside, and grip the door frame. Then the butterflies start. My heart starts thumping, too. It's silly, really. I've jumped free fall lots of times. But, then, I haven't ever done a water jump.

'Go', he shouts.

Too late to worry now! I push myself out of the door and start counting. 'One thousand, two thousand, three thousand', I shout, counting off the seconds. Arms spread, back arched – my position is good. I feel myself fall faster and faster.

'Fourteen thousand, fifteen thousand', I roar. Fifteen thousand – I've been falling for 15 seconds. I must be falling at 200 km per hour now. That's terminal velocity – I won't fall any faster.

'Nineteen thousand, twenty thousand'. Time to release the parachute. I pull the ripcord and the chute billows out. I don't like this part. There's always a powerful jerk when I slow down. It's the next part I like best – floating gently down. I know that I'm still falling at 20 km per hour, but it feels as if I'm suspended in mid-air.

There's the lake, shimmering below me. What a beautiful sight! Too bad I haven't time to admire the view. I have to land in the water. Hope I don't land far from the recovery boat.

Here comes the water. Time to inflate one side of the life jacket. Now to release the harness from my body and hang on tight to the straps. Body straight, arms stretched above my head and I'm in.

I'm sinking, but not very far. The water has slowed me down quicker than I expected. Time to start swimming. Half a dozen strokes and I'm clear of the harness, floating on the surface. What a great feeling – I've done it! Water jump number one.

Did you know?

- Some die-hard parachutists have been known to do water jumps with broken limbs still in plaster.
- In 1941, a Russian airman fell 6700 m without a parachute and survived. He landed on snow.

1 Joan's speed changes as she falls. What does the passage tell you about her speed: a) just after she leaves the plane b) after 15 seconds' freefall c) when her parachute opens?
2 How does gravity affect the speed at which she falls?
3 How does air resistance affect her as she falls?
4 Why does Joan slow down when the parachute opens?
5 Velocity is similar to speed. Work out what is meant by falling with terminal velocity, and why a free fall parachutist's speed does not keep increasing.
6 Friction in water is greater than in air. Which part of the passage tells you this?
7 **Try to find out:** how free fall parachutists control their speed.

13.4 Work and machines

When scientists talk about **doing work**, they use the words in a special way. They say that doing work is:

using a force to move an object

You use a force to push a mover, or pull a shoe lace. And so you are doing work when you mow the grass or tie your shoe.

Work is measured in **joules**. (You can see opposite how to calculate how much work is done.) The amount of work done depends on two things:

the size of force

the distance the object moves

Work, energy, and machines

You may remember having heard of a joule before. It is the unit used to measure energy. Work and energy are closely related! Your muscles need a supply of chemical energy if your body is doing to be able to do any work. An electric fork lift truck can do lots of work - as long as it has energy stored in its battery. But it can't do any work when the battery goes flat and the energy runs out.

If you have to do some hard work, a machine can be very useful. When you think of useful machines, you probably think of things like a crane, or an electric drill, or a washing machine. But for a scientist, machines can be much simpler than that. To a scientist:

a machine is anything which makes it easier to do work

And so a spanner is a machine. It makes it easier to do work. (Just try unscrewing a nut on your bicycle without one!) Opening a can, cutting wire, and lifting a car are other jobs which are impossible to do with your bare hands. But each job can be done quite easily using a simple machine - a can opener, pliers, a car jack ...

A machine makes it easier to do work for one main reason. Using a machine your can produce a much bigger force than you can on your own. Your fingers, for example, can't produce a big enough force to undo a nut. But if you use the same force on the end of a spanner, a far larger force is produced at the other end. This force is enough to make the nut turn.

How much work does the man do lifting the box?

Work done
= force × distance
= 120 N × 1.5 m
= 180 J

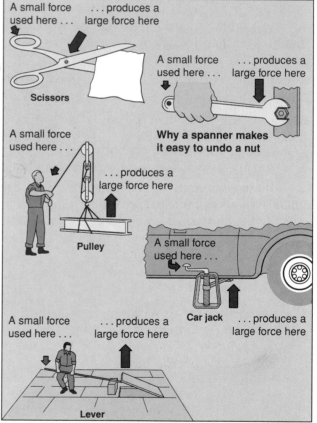

A small force used here . . . produces a large force here

Scissors

A small force used here . . . produces a large force here

Why a spanner makes it easy to undo a nut

A small force used here . . . . . . produces a large force here

Pulley

A small force used here . . .

Car jack . . . produces a large force here

A small force used here . . . . . . produces a large force here

Lever

1 What is meant by: a) doing work   b) a machine?
2 Are you doing work when you are: a) pushing a mower   b) cleaning your teeth   c) cycling   d) reading this page? Explain your answers.
3 Why must you eat if you are going to keep working?
4 Why are spanners, pliers, and can openers called machines?
5 **Try to find out:** the names of some other simple machines.

Did you know?

- Joule was an English scientist who lived between 1818 and 1899. He was the scientist who first showed that energy can be changed from one form to another.

Pulley wheels ...

With a set of pulley wheels and a rope, it is possible to make some difficult jobs much easier. A mechanics, for example, could never lift a car engine out of a car with his bare hands, but he could do it with a pulley system. Using the pulley, he can lift the engine **up** by pulling **down** on the rope. That's far easier than trying to lift the engine straight out of the car. He can also lift it using a far smaller force. The pulley has five lengths of rope supporting the engine and each rope supports it equally. If the engine weighs 500 N, each rope only needs to support 100 N, and that (roughly speaking) is that force which the mechanics has to use to lift the engine into the air.

...and bicycle wheels

A bicycle uses a kind of pulley system with a chain instead of a rope. When you push on the pedals, the chain **transmits** (passes on) the force to the back wheels, turning them.

The cycle's gears change the number of times that one turn of the crankshaft turns the back wheels. When you are riding along a flat road, you put the bike in high gear. Then, the chain runs over the small cog wheel at the back so that one turn of the crank turns the back wheel several times. The force you can put on the pedals is enough to do this. The bike speeds along.

When you are climbing a hill, however, things are much more difficult. The bike has to be in low gear. The chain now has to run over the large cog wheel at the back. One turn of the crank only turns the back wheel once or twice, but this means that the force you use will be enough to keep the bike going uphill.

High gear – the chain runs over the larger gear wheel on the crank and the smallest gear wheel at the back. (One turn of the pedals turns the back wheel several times.)

1 What do you need to make a pulley system? ▲
2 Give two reasons why it is easier for a mechanic to lift a car engine out using a pulley system. ▲
3 What is the job of: a) the chain b) the gears on a bicycle?
4 Turning a bicycle crankshaft turns the back wheel. What is the different between cycling in high gear and low gear?
5 **Try to find out:** some other uses of pulleys.

Did you know?

● The earliest cycles had wooden frames and wheels and iron tyres. They were called **boneshakers**. It's not difficult to understand why!

81

Going further

The Great Pyramid of Egypt is an impressive building, even by modern day standards. Rising to a height of 140 m, it is built from 2 300 000 blocks of stone, each weighing 25 000 N or more. It's hard to believe that the Egyptians built it over 5000 years ago using only the simplest of machines like wedges, levers, ramps, and rollers. But, then, they did find 100 000 slaves a big help!

The blocks used in the Pyramid were cut out of quarries on both sides of the river Nile. They were moved over the ground on rollers.

Flat-bottomed, barge-like boats were used to transport the blocks across the river. These were towed by larger boats which were driven by sails and oars.

To move the blocks into place, the slaves had to pull them up huge ramps of earth. Of course they had to build the ramps first of all.

Wedges were used to split the blocks. Holes were cut in the rock and wooden wedges were hammered in. When water was poured over the wedges the wood swelled, splitting the stone.

Levers were used to raise the blocks onto the sledges and rollers. The rudder which steered the ship was also a lever. So were the oars.

The ramps and rollers had to be used because the Egyptians had no cranes to lift the blocks into place. The heaviest block weighed 150 000 N.

1 Which simple machines were used: a) to split the blocks
   b) to move the blocks   c) to put the blocks in place? ▲
2 a) How was friction reduced when the blocks were moved? ▲
   b) A simple invention, first used 700 years after the Pyramid was built, would have reduced friction further. What is it?
3 Draw a diagram to show how to lift a block of stone using a lever.
4 Wedge shaped tools and ramps are still in use. Give some examples of how they are used.
5 How much work had to be done to raise the last block (weighing 25 000 N) on to the top of the Pyramid?

Did you know?

● The blocks of stone were so accurately cut that the gaps between blocks in the Pyramid are 0.5 mm or less.
● The most powerful crane in the world can lift 350 of the heaviest blocks.

13.4 Wasted work

Some rather unusual racing bicycles appeared in the 1984 Olympic Games. Their frames had an unusual shape and were made of aluminium, not the normal steel. They had narrower hubs than normal, and smaller wheels. These back wheels were solid, and the tyres were filled with helium, not air. These **'funny bikes'** had been designed for the U.S.A. team with one purpose in mind – to make sure that their cyclists wasted as little work as possible!

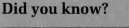

A 'funny bike'

Wasting work – it can't be helped!

Every cyclist wastes work. To make a bicycle go, a cyclist has to do work, pushing the pedals round. Some of this work is useful. It moves the cyclist forward. But some work has to be done to move the bicycle forward, and to overcome friction. That's wasted work!

The 'funny bike' designers:

Made the bicycle as light as possible to cut down the amount of work which has to be done to move the cycle forward. Using aluminium instead of steel cut the weight by 28%. Using helium instead of air cut the weight by 0.3% more!

Cut down friction. Using the odd-shaped frame and flattened spokes cut down air resistance. And, of course, the moving parts were well oiled.

All of this cut 1 second off a rider's time in a 1000 m race!

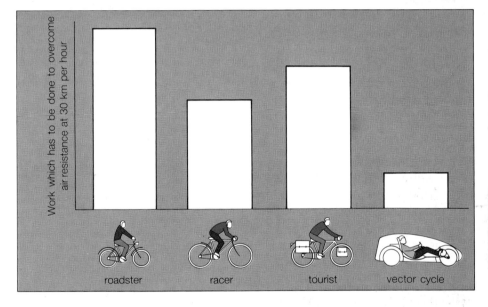

It's the same for every machine!

Like the bicycle, every machine wastes some of the work put into it. A car jack, for example, wastes 60% of the work you do when you turn the handle. There is a lot of friction between its moving parts.

The jack is said to be **40% efficient** because only 40% of your work is actually converted to useful work, raising the car. Many other machines are more efficient than this. A pulley, for example, can be 75% efficient, only wasting 25% of the work you do.

1 Some of the work done by a cyclist is useful work. What does this work do? Why is the rest of his work said to be wasted work? ▲
2 How did the designers of the 'funny bikes' cut down the amount of work wasted by the cyclists? ▲
3 A well lubricated machine wastes less work. Explain why.
4 What is meant by saying, 'a car jack is 40% efficient'. ▲
5 Which cyclist in the diagram is most affected by air resistance? Suggest why air resistance has such a large effect on him.

Did you know?

- This is one of the most efficient bicycles in the world. It once travelled at 62.9 kph.
- The U.S. team wore crash helmets which were specially designed to cut down air resistance.

14.1   What's in a food?

Going further

Most people think that bread is made of carbohydrate and they're correct – up to a point. Bread does contain lots of carbohydrate, but it also contains protein, fat, vitamins, minerals, and water.

Most foods, like bread, are mixtures, but different foods are very different mixtures! Some foods are mostly made up of carbohydrates, some of fat and so on. The colour code diagrams show that!

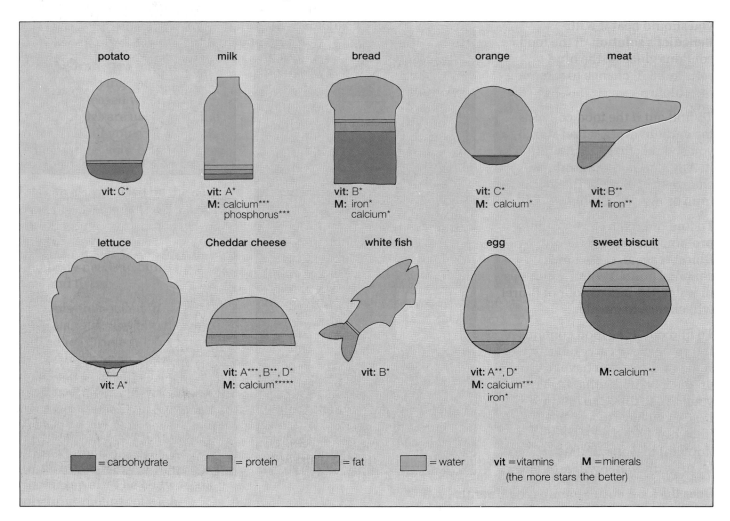

Questions 1–5 are about the foods in the colour-code diagrams.

1 Which foods give the best supplies of: a) protein   b) vitamins
   c) minerals?
2 Which foods are mostly made up of: a) carbohydrate   b) water?
3 If a doctor told you not to eat fats, which of the foods should you cut
   out?
4 a) Why can you eat lots of green vegetables when you are slimming?
   b) Which of the foods should you only eat in small quantities when
   on a diet?
5 Cheddar cheese is made from milk. What differences are there
   between the two?
6 **Try to find out:** what chocolate and butter are made of. Then draw
   colour code diagrams with the information.

Did you know?

- There is as much protein in peanuts as in meat. Soya beans contain more protein than either.
- There is more water in an apple than in milk.

90

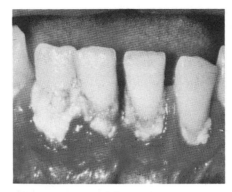

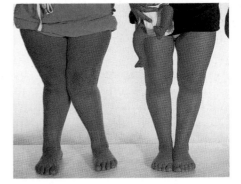

Night-blindness is a disease which prevents people from seeing properly in dim light. Ancient Egyptian 'doctors' knew how to cure it. They suggested that, if a patient ate ox liver, he would be cured. This is a statue of Imhotep, the father of Egyptian medicine.

Scurvy causes bleeding gums, muscle weakness and death. It was a disease which, for many years, affected sailors on long voyages. In 1747, Dr James Lind found that drinking fruit juice could cure it. Later, ocean-going British ships carried limes to treat the illness.

Rickets affects bones. It makes them soft. Then the bones can become bent and twisted. Even as late as 1940, some British children suffered from the disease. But long before that, doctors knew that it could be cured with cod liver oil.

These doctors did not know what had caused the diseases, but we now know that they are caused by shortages of vitamins in the diet. The doctors' work did, however, help scientists to find out more about vitamins. But, as the story of vitamin B_1 shows, finding out everything about a vitamin can take a long time. Vitamins are complicated chemicals, and are only found in tiny amounts.

The story of vitamin B_1 (the first vitamin to be found)
1896 A Dutch doctor, Dr Eijkman, went to Java looking for a cure for a disease called **beri-beri**. He noticed that some hospital chickens had a similar disease. Then, one day, they started to recover. Dr Eijkman found that their diet had changed from polished rice (rice with no husks) which had been thrown out by the hospital, to whole rice (rice with husks).
1901 Eijkman's colleague, Dr Grijns, decided that rice husks contained a chemical which cured beri-beri.
1906 Eijkman and Grijns boiled rice husks in water. This water cured a pigeon suffering from a disease like beri-beri.
1912 Chemists extracted the curing 'chemical' from rice husks.
1934 The chemical was identified. It was called vitamin B_1.
1937 Chemists made vitamin B_1 in the laboratory.

1   Why are vitamins: a) necessary   b) difficult to investigate? ▲
2   Which 2 pieces of evidence suggest that liver contains vitamins?
3   Why does rice lose its vitamin B_1 when polished?
4   Why did a change of diet cure the hospital chickens?
5   Why are modern sailors not affected by the vitamin shortage diseases which affected sailors 300 years ago?
6   Which vitamin prevents scurvy? (Information on page 90 will help.)
7   **Try to find out:** why liver cures night blindness, and why cod liver oil and sunshine can help to cure rickets.

In Asian countries, polished rice makes up a large part of the diet

Did you know?

● British sailors are nick-named 'Limeys'.
● Most vitamins are now made by the chemical industry, not extracted from living things.
● There are some vitamins which can do you harm – if you eat too much of them.

14.2   Breaking down food [1]: teeth

Food has to be broken down into small pieces before your body can use it. You can't swallow an apple whole! To eat the apple (or any large piece of food), you first have to bite off a piece of it. Then you have to chew the piece until it has been ground down enough for easy swallowing. Muscles, jaws, and teeth all play a part in this breakdown of the food. The muscles supply the movement. They pull on the jaws and keep them moving. Since the teeth are firmly fixed in the jaws, this keeps the teeth moving, biting, and grinding.

When an adult laughs, this is what you should see – a set of 32 gleaming white teeth. (Your smile will be less toothy! You can't expect to have 32 teeth until you are about 18).

Each jaw has 16 teeth in it. The front 4 are sharp biting teeth, called **incisors**. Behind them are 2 **canine** teeth also used for biting. The other 10 'back' teeth are much flatter. Their job is to grind the food into tiny bits. Four of these are **premolars**. The other six are **molars**.

The part of the tooth which you can see is covered by a layer of white **enamel**. This is a very hard, non-living substance. It protects the tooth and prevents it from being worn away.

The enamel covers a living part of the tooth which is made of **dentine**. The dentine is softer than enamel and is a bit like bone. In the centre of each tooth is the **pulp cavity**. This is made of soft pulp, which is made up of living cells. It also contains nerves and blood vessels.

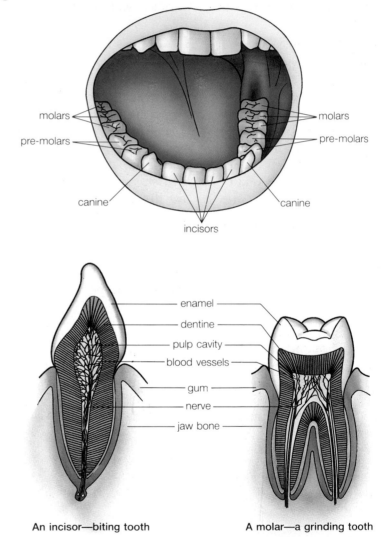

An incisor—biting tooth   A molar—a grinding tooth

1   What part do: a) teeth b) jaws c) muscles play in breaking down food. ▲
2   What job is carried out by: a) the incisors b) the molars? ▲
3   What is: a) dentine b) enamel? ▲
4   The enamel protects the tooth. Why is it suitable for this?
5   How many incisors, canines, and molars do you have in each jaw? (Ask a friend to help if you find if difficult to count!)
6   **Try to find out:** why canine teeth got their name.

Did you know?

● Enamel is the hardest substance in the human body, but it is affected by acid.

14.2   The story of your teeth?

The story of an adult tooth always begins in the same way. The tooth grows in the jaw underneath the gum. As it grows, it is gradually pushed towards the surface. By the time this happens, the baby teeth are well worn down. Their roots get smaller and smaller until each baby tooth falls out. Then the adult tooth pushes through to fill the gap.

How the story ends for your teeth depends on you! It can have a sad ending, or a happy one.

If you don't take care, your tooth will come to a sticky end!

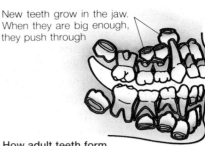

New teeth grow in the jaw. When they are big enough, they push through

How adult teeth form

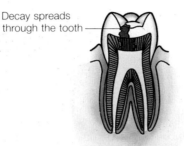

Decay spreads through the tooth

1 When you eat a food, a sticky substance called **plaque** forms on your teeth. Plaque has bacteria in it. If you eat sweet food, lots of bacteria grow.

2 The bacteria change sugary food to acid. The more sweets you eat, the more acid is made. The acid eats through the enamel, making a hole. Decay starts to spread.

3 A dentist can stop the rot. If you go to him quickly enough, he can fill the hole. But if the decay has gone too far, he will only be able to pull out the tooth.

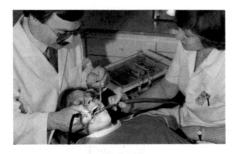

1 If you cut down on sweet food, less plaque will form. If you brush your teeth well, much of the plaque which does form will be brushed away.

2 If you visit your dentist often, he will keep a close check on your teeth. He will fill holes as soon as they appear.

3 Then your smile at 40 won't be very different from your smile at 14! And you won't have to keep your teeth in a glass overnight!

1 How old were you when you lost your first 'baby' tooth? Which tooth was it? Why did it fall out?
2 a) What is plaque, and how does it form? ▲
    b) Why does plaque cause problems for your teeth? ▲
3 a) Give some good reasons for going to the dentist regularly.
    b) When did you last go to the dentist? What did he (or she) do?
4 You can't feel enamel decaying. Suggest why not.
5 Make a poster with some rules about caring for your teeth.
6 **Try to find out:** about fluoridation of drinking water.

Did you know?

- Fluoride toothpaste helps to prevent tooth decay. Fluoride hardens the enamel.
- False teeth have been used since 1000 BC. The Ancient Greeks used to tie them to their good teeth using strings and wires.

Dogs, cats, and other **carnivores** (meat eaters) have teeth which are ideal for the way they feed. They have long canines with which they can stab their prey and prevent it from escaping. They have sharp pointed molars which allow them to slice large pieces of meat off bones, and to break up bones to get the marrow inside. They have small sharp incisors which are useful for cutting away flesh next to the bone. These animals don't grind their food. They swallow it in lumps.

The teeth of **herbivores** (plant eaters) are ideal for cutting and chewing grass and other vegetation. If you look at the diagram of the sheep's teeth, you will see that, like many herbivores, the sheep has a horny pad instead of upper incisors. When the sheep bites off a piece of grass, it takes the grass between this pad and its incisors. It cuts the grass by pushing its jaw forward. Then it has to grind the grass very thoroughly to break up the tough plant cell walls. Its large, close fitting molars allow it to do this.

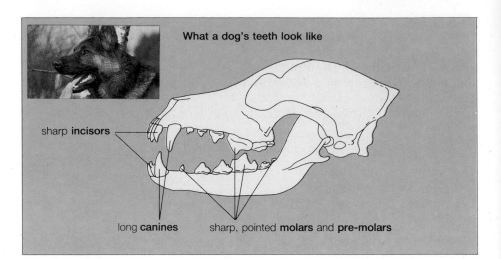

What a dog's teeth look like

sharp **incisors**

long **canines**　　sharp, pointed **molars** and **pre-molars**

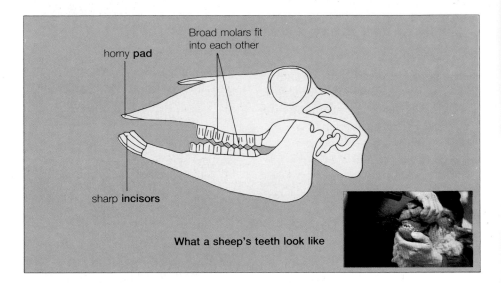

Broad molars fit into each other

horny **pad**

sharp **incisors**

What a sheep's teeth look like

Did you know?

● You are an **omnivore**. You can eat any kind of food. Your teeth are suitable for eating meat or plants.
● Carnivores' jaws can only move up and down. Their teeth shear meat with a scissors-like action. Herbivores' jaws can move sideways, which helps the grinding down of their food.

1　What is:　a) a carnivore　b) a herbivore? ▲
2　When a dog is eating a bone, which teeth does it use:　a) to get flesh off the bone　b) to crack the bone open? ▲
3　How are the molars of a carnivore and a herbivore different in shape? How is each type ideal for the job it has to do? ▲
4　What is a cow doing when it is 'chewing the cud'?
5　Look at the wild animal's skull. Then suggest:　a) if the animal was a herbivore or a carnivore (giving reasons)　b) the animal's name.
6　**Try to find out:** which animals have teeth like yours.

This skull, of a British wild animal, was found on a moor. The skull is around 12 cm long

14.3   Breaking down food [2]: digestion

Food has to get into the blood to be carried to the body's cells. Only **soluble** food (food which dissolves) can do this.

Most of the food you eat, however, is **insoluble**. Even if you grind it down finely, it still won't dissolve. And so, to make use of it, your body has to break it down into chemicals which can dissolve. This breakdown is called **digestion**. It takes place in the **digestive system**.

Breaking down the food is the job of your **digestive juices**. The breakdown of some food starts in your mouth. There the food is mixed with a juice called **saliva** which is made in your **salivary glands**. As the food passes through the digestive system, other juices are added from, for example, the **liver** and the **pancreas**. Further breakdown of the food takes place.

Muscles also have an important part to play in digestion. Muscles keep the walls of the stomach and small intestine moving. This mixes the food and digestive juices inside, speeding up digestion. Muscles also keep the food moving through the digestive system.

When the food has been completely broken down, it is **absorbed** into the blood. This happens in the last part of the **small intestine**. It has a good supply of blood and thin walls. This allows food to pass easily into the blood.

Some of the food you eat can't be digested. Your body gets rid of this waste through the anus.

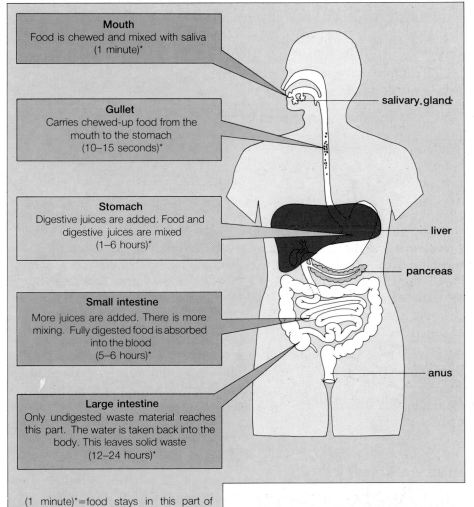

Mouth
Food is chewed and mixed with saliva
(1 minute)*

Gullet
Carries chewed-up food from the mouth to the stomach
(10–15 seconds)*

Stomach
Digestive juices are added. Food and digestive juices are mixed
(1–6 hours)*

Small intestine
More juices are added. There is more mixing. Fully digested food is absorbed into the blood
(5–6 hours)*

Large intestine
Only undigested waste material reaches this part. The water is taken back into the body. This leaves solid waste
(12–24 hours)*

salivary gland
liver
pancreas
anus

(1 minute)*=food stays in this part of the digestive system for around 1 minute

1 What happens to food in digestion? Why is this important? ▲
2 What job is done by digestive juices? Where are they made? ▲
3 What part do muscles play in digestion? ▲
4 a) Bread is mostly made up of carbohydrate. If you ate toast for breakfast, where could the carbohydrate be now?
   b) Some of the carbohydrate will be used by cells in your feet. Describe how it gets from your mouth to your feet.
5 a) What happens to fully digested food in the small intestine? ▲
   b) **Try to find out:** why the wall of the intestine allows this to happen easily.

Did you know?

- Chewing does matter! Finely ground food is digested more quickly than large lumps.
- Carbohydrates spend the shortest time in the stomach. Fats spend the longest.
- An adult's digestive system is about 10 m long.

14.3 Breaking up molecules

Glucose and starch are two carbohydrates. They are both made up of molecules, but the molecules are very different. Glucose molecules are small . . . small enough to dissolve, and small enough to get through the wall of the intestine. And so, when you eat a glucose tablet, you don't have to digest it. The glucose dissolves in your mouth. When it reaches your intestine, it is quickly absorbed into the blood.

When you eat a starchy roll, however, it's a different story. Compared with glucose molecules, starch molecules are huge. They are far too big to dissolve, and far too big to get through the intestine wall. And so starch has to be digested before your body can use it.

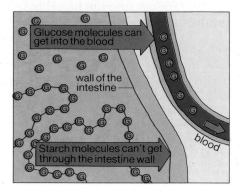

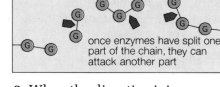

Enzymes at work

How does digestion change starch?

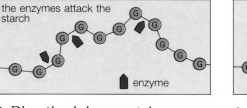

1 A starch molecule is really made up of glucose. It's a long chain made of many glucose molecules joined together.

2 Digestive juices contain chemicals called **enzymes**. Some of these enzymes can attack the starch chain and can split it.

3 When the digestive juices are mixed with starch, the starch is broken down to glucose. This glucose can then be absorbed into the blood.

The same kind of thing happens when proteins and fats are digested. Both of these chemicals are made up of large molecules which can't get through the intestine wall. Both can be broken down, by enzymes, into smaller molecules. When your food is churning around in your stomach, and intestine, the enzymes are hard at work. They're breaking up molecules.

Did you know?

- Each enzyme can only do one job. Enzymes which break down starch can't break down proteins or fats.
- Enzymes aren't only involved in digestion. They control many of the processes going on in your body, and in other living things.

1 Glucose molecules don't have to be digested. Starch molecules do. Explain the difference. ▲
2 What are digestive enzymes? What do they do? ▲
3 What happens to starch when it is digested? ▲
4 Saliva only contains one enzyme. That enzyme breaks down starch. Suggest why: a) bread begins to taste sweet when you chew it for a long time   b) the digestion of meat does not start in the mouth.
5 **Try to find out:**   a) why biological detergents use enzymes
   b) why herbivores can use grass as a food, but humans can't.

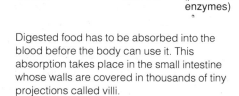

Digested food has to be absorbed into the blood before the body can use it. This absorption takes place in the small intestine whose walls are covered in thousands of tiny projections called villi.

96

14.3 A closer look at enzymes

Do enzymes work faster as the temperature goes up?
That's the question which Kate's teacher asked the class to answer.
He gave the pupils starch solution, enzyme solution, and iodine solution.
He asked the pupils to carry out an experiment to find if enzymes work faster at some temperatures than at others.

Kate carefully planned out what she was going to do. (There seemed to be lots of things to consider.) Then she carried out the experiment in three steps.

Step 1 She took 6 test tubes and put 10 cm^3 of starch solution and 1 cm^3 of enzyme solution in each

Step 2 She put the test tubes into beakers of water at different temperatures, then kept the temperature steady.

Step 3 Every 10 minutes, she tested the solution. She took a sample from each test tube, added it to iodine on a spotting tile, and noted the colour.

Here are Kate's results. Can you make sense of them?

Test Tube	Temperature of water (°C)	Colour of mixture at— (time, min)							
		10	20	30	40	50	60	70	80
1	10	●	●	●	●	●	●	●	●
2	20	●	●	●	●	●	●	●	●
3	30	●	●	●	●	●	●	●	●
4	40	●	●	●	●	●	●	●	●
5	50	●	●	●	●	●	●	●	●
6	100	●	●	●	●	●	●	●	●

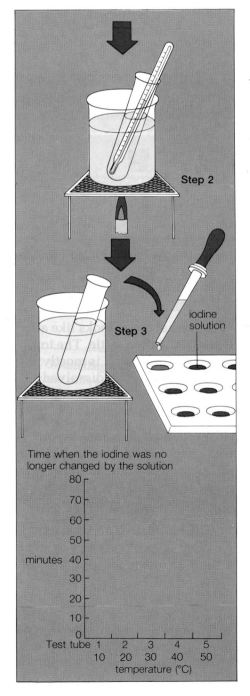

Step 2

iodine solution

Step 3

Time when the iodine was no longer changed by the solution

1 Why did Kate add the same amount of enzyme to each test-tube?
2 What does saliva do to starch? (see page 96 Q4)
3 Why did each of the solutions turn the iodine blue-black at the beginning? Why did some of the solutions not affect the iodine after a time?
4 How could Kate tell how quickly the enzyme affected the starch?
5 Draw and complete the bar chart opposite. Then use it to work out an answer to Kate's teacher's question.
6 From the results, work out what boiling does to enzymes.
7 **Try to find out:** a) the temperature at which enzymes work best
b) why vegetables are quickly boiled before freezing.

What on earth does it say?

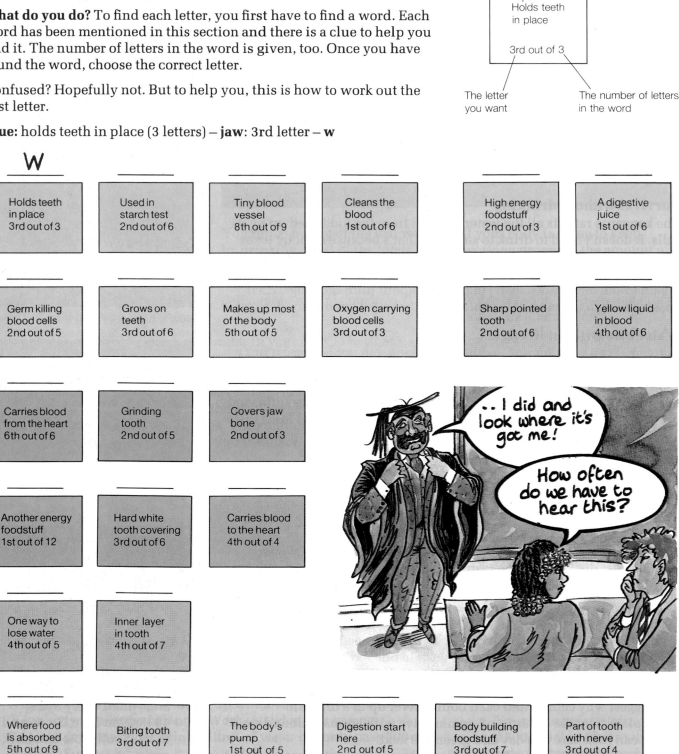

Here is a piece of light relief to finish off this section. It's a coded message, containing a piece of advice. Many (older!) folk would say that it's worthwhile working out what it says. They think it's the secret of success.

What do you do? To find each letter, you first have to find a word. Each word has been mentioned in this section and there is a clue to help you find it. The number of letters in the word is given, too. Once you have found the word, choose the correct letter.

Confused? Hopefully not. But to help you, this is how to work out the first letter.

The clue to the word

Holds teeth in place

3rd out of 3

The letter you want

The number of letters in the word

Clue: holds teeth in place (3 letters) – **jaw**: 3rd letter – **w**

W

| Holds teeth in place 3rd out of 3 | Used in starch test 2nd out of 6 | Tiny blood vessel 8th out of 9 | Cleans the blood 1st out of 6 | High energy foodstuff 2nd out of 3 | A digestive juice 1st out of 6 |

| Germ killing blood cells 2nd out of 5 | Grows on teeth 3rd out of 6 | Makes up most of the body 5th out of 5 | Oxygen carrying blood cells 3rd out of 3 | Sharp pointed tooth 2nd out of 6 | Yellow liquid in blood 4th out of 6 |

| Carries blood from the heart 6th out of 6 | Grinding tooth 2nd out of 5 | Covers jaw bone 2nd out of 3 |

| Another energy foodstuff 1st out of 12 | Hard white tooth covering 3rd out of 6 | Carries blood to the heart 4th out of 4 |

| One way to lose water 4th out of 5 | Inner layer in tooth 4th out of 7 |

.. I did and look where it's got me!

How often do we have to hear this?

| Where food is absorbed 5th out of 9 | Biting tooth 3rd out of 7 | The body's pump 1st out of 5 | Digestion start here 2nd out of 5 | Body building foodstuff 3rd out of 7 | Part of tooth with nerve 3rd out of 4 |

106

Electricity in action

Have you ever been stuck at home . . .

. . . during a power cut . . .

. . . at night . . .

. . . in winter?

Life's not much fun without electricity.

No television.

No record players.

No lights.

No electric cookers.

No electric heaters.

You only realise how much you depend on electricity when the power goes on again.

There's a lot to learn about electricity. This section will teach you a little more about it. In the section you will find out how electricity is generated, and how it is used. But first you will learn a little about magnets. Magnets have an important part to play when **electricity is in action**.

15.1 Permanent magnets

You already know a little about magnetism. It's one of the forces which you met on page 70. It's a force which can help you if you spill a box of pins. But it can't do much to help if you spill a box of matches.

A magnet can attract anything which has iron, cobalt, or nickel in it. It picks up pins because they are made of steel, which is mostly iron. The magnet doesn't have to touch each pin to pick it up. The pins jump across the gap when the magnet is brought up close to them. That's because:

a magnet affects the space around it.

Once the tin has been opened, a magnet lifts the lid off

The space around a magnet in which a magnetic force acts is called the **magnetic field**. The magnetic field is strongest at the magnet's **poles**. That's where the greatest magnetic force acts. The further away from the magnet, the weaker the field becomes.

Each magnet has two poles, a **north pole** and a **south pole**. If the north poles of two magnets are brought together, the magnets **repel** each other (push each other away). The same thing happens when two south poles are brought together. But when a north pole and a south pole are brought together, they **attract** each other.

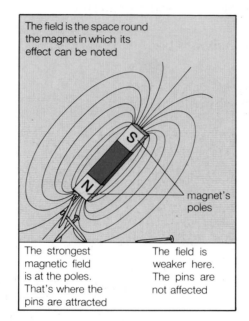

The field is the space round the magnet in which its effect can be noted

magnet's poles

The strongest magnetic field is at the poles. That's where the pins are attracted

The field is weaker here. The pins are not affected

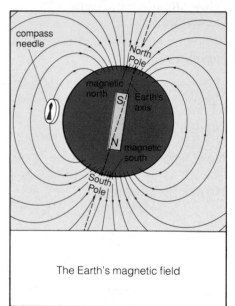

compass needle

North Pole

magnetic north

Earth's axis

magnetic south

South Pole

The Earth's magnetic field

The Earth – a giant magnet?

The Earth has a magnetic field. It behaves as if it has a giant bar magnet inside it.

It's the Earth's magnetic field which allows you to use a compass to find directions. Whenever a bar magnet is able to swing freely, this magnetic field makes it line up pointing in the same direction (roughly north and south). A compass needle is a small bar magnet. That's why it always points north and south.

1   Why can a magnet attract pins, but not matches?
2   What is a magnetic field? ▲
3   Where is the field round a magnet: a) strong   b) weak? ▲
4   When will two bar magnets: a) repel   b) attract each other? ▲
5   a) What makes a compass needle always point in the same direction? ▲
    b) The Cuillin Hills in Skye contain some magnetic rocks. Why could this cause problems for mountaineers?
6   **Try to find out:**   a) how to magnetise a nail   b) what the first compasses were made of.

Did you know?

- Magnets can be made of iron, cobalt, or nickel.
- Powerful magnets affect digital watches.
- The magnets described on this page are permanent magnets. They keep their magnetism unless you destroy it, by heating, for example.

15.1 Electromagnetism

Around the beginning of the 1800s, there was lots of interest in electricity. The first battery had just been invented. Electricity from batteries was used to carry out all sorts of experiments.

In 1820, a Danish scientist called Oersted was giving a lecture in Copenhagen. In it, he carried out an experiment to show (he thought) that there was no connection between electricity and magnetism. He put a compass needle and a wire side by side. Then he passed an electric current through the wire. Imagine his surprise – and embarrassment – when the compass needle swung round!

What Oersted had found – by accident – was a really important discovery:

When a current flows through a wire, the wire has a magnetic field round it.

The compass needle was a small magnet. It was affected by the magnetic field round the wire. That's why it swung round.

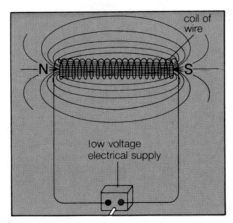

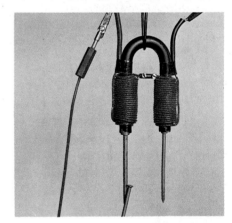

When a current flows through a single wire, the magnetic field round it is very weak. If the wire is made into a coil the field is stronger. Putting a piece of iron inside the coil makes the field even stronger!

A coil of wire behaves like a bar magnet when a current flows through it. One end of the coil behaves like a magnet's north pole and the other like a south pole. Changing the direction of the current changes round the poles.

Most electromagnets are made by coiling wire round cores made of soft iron. When a current passes through the coil, the core behaves like a magnet. Switching off the current destroys the magnetism.

Did you know?

- You can magnetise a piece of iron by putting it inside a coil of wire and passing a current through the wire.
- A little copper is mixed with the iron of an electromagnet. This makes the magnet lose its magnetism when the current is switched off.

1 What affected the compass needle in Oersted's experiment? ▲
2 A coil of wire behaves like a magnet when a current flows through it. What could you do to make the magnetic effect stronger? ▲
3 What would you need to make an electromagnet? ▲ Explain how you would control it once you have made it.
4 Why does an electromagnet have copper in it? ▲
5 Look back at *Did you know?* on page 108. Then write down the main difference between electromagnetism and permanent magnetism.
6 **Try to find out:** more about Oersted.

Going further

Electromagnets are used to do many different jobs.

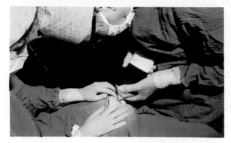

Huge electromagnets are used in **scrapyard cranes**. They are powerful enough to lift cars. They are also useful for sorting out scrap iron from other metals.

In hospital, a surgeon can use an electromagnet to remove a piece of iron from a patient's eye.

This is **Maglev**, a train which runs at Birmingham airport. When it is running, electromagnets lift it clear of the track. This gives a smooth, quiet, friction-free ride.

Some **burglar alarms** also use electromagnets. The alarm system shown below is simpler than a real one, but it does give you an idea about the job which an electromagnet can do.

This is an alarm for a window. It has two metal contacts fitted into the window frame.

As long as the window is shut, these contacts touch. A current flows in circuit 1 and so the electromagnet attracts the piece of soft iron. There is a gap in circuit 2 and so no current flows.

As soon as the window is opened, the contacts spring apart. Circuit 1 is broken, switching off the electromagnet. The spring pushes the piece of iron across to complete circuit 2. This allows a current to flow through the bell, which rings.

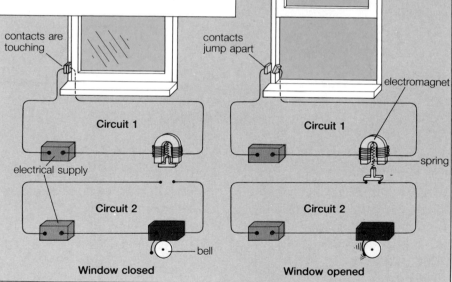

1   Why are electromagnets useful: a) in scrapyards
    b) in hospitals? ▲
2   Imagine you were a crane driver in a scrapyard. Explain exactly what you would do to lift a car across the yard with an electromagnet.
3   Look at the burglar system above. Explain:
    a) why the bell does not ring all of the time
    b) why the bell starts to ring when the window is opened
    c) what would happen in a power cut.
4   A normal train can be affected by icing of the track, but this does not bother Maglev. Explain the difference.
5   **Try to find out:** how an alarm bell works.

Did you know?

- A special kind of motor, which uses electricity and magnetism, drives Maglev forwards.
- Computers on Maglev control the height at which it rides. By altering the power of the electromagnets, they keep it 15 mm above the track.

If you had to build a big, strong electromagnet, what would you do? Which material would you use to make the core? How many turns of wire would you wind round it? What size of current would you use? All of these things matter.

Here are the results of some experiments. They will help you work out the answers to these questions. The electromagnets in the diagrams are all different. The number of paper clips which they can lift varies, too. You can tell how strong the magnet is from the number of paper clips it supports.

Two things to notice:

1 The electromagnet cores are all the same size.
2 When the current is switched off, the paper clips drop from the iron-cored magnet, but not from the steel-cored one.

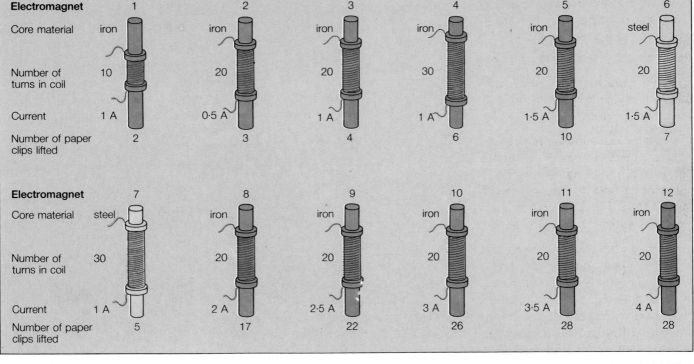

Electromagnet	1	2	3	4	5	6
Core material	iron	iron	iron	iron	iron	steel
Number of turns in coil	10	20	20	30	20	20
Current	1 A	0·5 A	1 A	1 A	1·5 A	1·5 A
Number of paper clips lifted	2	3	4	6	10	7

Electromagnet	7	8	9	10	11	12
Core material	steel	iron	iron	iron	iron	iron
Number of turns in coil	30	20	20	20	20	20
Current	1 A	2 A	2·5 A	3 A	3·5 A	4 A
Number of paper clips lifted	5	17	22	26	28	28

1 Write down three things which could affect the strength of an electromagnet. ▲
2 In this experiment, how can you tell how strong the magnet is? ▲
3 If you wanted to find out how the number of turns affects the strength, which three magnets would you compare? Why did you choose them? How does the number of turns affect the strength?
4 a) Which magnets would allow you to find out how the strength of a magnet depends on the current?
b) Draw and complete the graph on the right for iron-coiled magnets only with 20 turns of coil.
c) 'The bigger the current, the stronger the electromagnet.' Is this true?
5 Iron is a better core material than steel. Give one reason why.

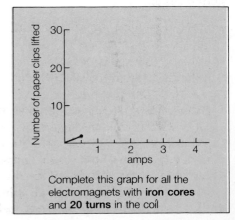

Complete this graph for all the electromagnets with **iron cores** and **20 turns** in the coil

15.2 Generating electricity

You have just found out that electricity can be used to make magnets. The opposite is also true. Magnets can be used to produce, or **generate**, electricity.

It's quite easy to generate electricity in the lab. All you need is a magnet (a U-shaped one is best) and a loop of wire.

To generate the electricity, you have to supply moving energy. You must move the wire up and down between the magnet's poles.

To show that a current is flowing you have to connect in a meter. It will give a tiny reading, but only while the wire is moving.

This experiment shows that: **a current flows while a conducting wire is being moved through a magnetic field.**

Useful electricity can be generated in the science lab, but not with a single wire moved by hand! To do this, you should:

- wind a long wire into a coil
- mount the coil on an axle
- place the coil between the poles of a magnet
- spin the coil steadily.

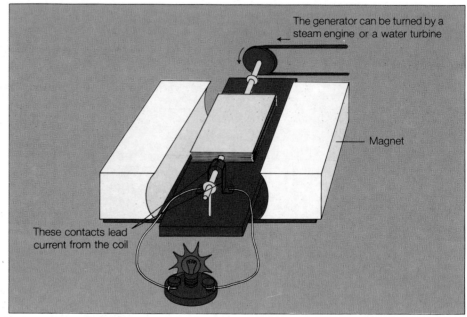

The generator can be turned by a steam engine or a water turbine

Magnet

These contacts lead current from the coil

This is the arrangement in the model **generator** shown above. It uses the moving energy from a steam engine or a water turbine to spin the coil. When it is working, this generator can supply a steady current big enough to light a torch bulb. A power station generator is much more complicated and much more powerful. It can generate enough electricity to supply a whole town!

1 What is a generator's job?
2 a) How can you generate electricity using a wire and a magnet? ▲
   b) What is the energy change when you do this?
3 How would you make a model generator? ▲
4 What is the energy change when a generator is driven by:
   a) a windmill   b) a diesel engine   c) a steam engine burning meths.
5 a) Why must hospitals have stand-by generators?
   b) **Try to find out:** where else stand-by generators are used.

Did you know?

- The girl is producing **alternating current, a.c.** This kind of current keeps changing direction, flowing along the wire in one direction, then back again.
- Hospitals have stand-by generators in case of power failure. They are driven by diesel engines.

Spinning a coil between the poles of a fixed magnet isn't the only way to generate electricity. Spinning a magnet inside a fixed coil generates electricity just as well.

The **bicycle dynamo** in the photograph uses a spinning magnet. You have to supply the energy to spin it. To make the dynamo generate electricity, you have to press its driving wheel against the bicycle's tyre. To make the wheel turn, you have to pedal!

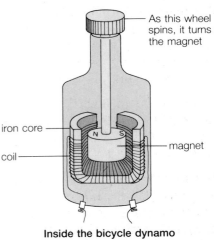

As this wheel spins, it turns the magnet

iron core

coil

magnet

Inside the bicycle dynamo
(a cut-away drawing)

This explains why your bicycle headlamp gets brighter and dimmer while you are cycling in the dark. The current generated by the dynamo depends on how fast your bicycle wheel turns, as you can see from the diagram below.

A winter's night cycle ride home from school

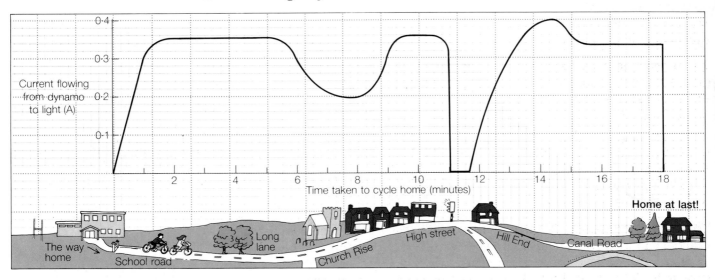

Current flowing from dynamo to light (A)

Time taken to cycle home (minutes)

Home at last!

The way home　School road　Long lane　Church Rise　High street　Hill End　Canal Road

1　How does a bicycle dynamo work?　▲
2　With a bicycle, a dynamo, and a lamp, you can change chemical energy in your food into light energy. How does this happen?
3　Why is it more difficult to pedal when the headlamp is lit?
　　About the cycle ride home from school
4　Where was the bicycle after:　a) 2 min　b) 7 min　c) 16 min?
5　Why was the headlamp brighter at some times than at others?
6　What was the biggest current generated by the dynamo?
7　Suggest why the current dropped:　a) on Church Rise
　　b) on High Street.
8　For the rider to see clearly, the lamp must be supplied with 0.3 A or more. For how long was this current supplied?

Did you know?

● Using a spinning magnet to generate electicity has an advantage over using a spinning coil. It's easier to take current from a fixed coil than from a moving one.

A power station generator is called an **alternator**. It looks like this . . .

. . . from the side

. . . from the front

The power station alternator generates electricity in the same way as the bicycle dynamo. The electricity is produced by spinning a magnet inside fixed coils of wire. But that's where the resemblance ends! The power station alternator:

. . . is much bigger. A typical alternator is about 65 m long, 6 m high and has thousands of turns of wire in each coil.

. . . generates a far larger current – perhaps 20 000 A at a voltage of 25 000 V.

. . . has a cooling system. The currents produce large amounts of heat when they flow through the coils. This could cause damage, and so the alternator has to be cooled. In some alternators, water is piped through the coils to cool them.

. . . uses a spinning electromagnet instead of a permanent magnet.

. . . has the magnet spinning at a steady rate. British power stations supply **alternating current (a.c.)** with a **frequency of 50 hertz**. Fifty times each second, the alternators push the current one way along the wire and pull it back again. To achieve this, the electromagnet must spin exactly 50 times a second. The spinning speed must be steady. Most electrical appliances have been designed to use a current with a frequency of 50 Hz, and won't work properly if the frequency changes.

The electricity is generated in these coils

spinning electromagnet

1   What is an alternator? What is its job? ▲
2   How are the bicycle dynamo and the power station alternator: a) like each other   b) different from each other? ▲
3   a) Why does a power station alternator have a cooling system? ▲
    b) Why is the alternator filled with hydrogen? ▲
4   a) How often does an alternator's electromagnet spin in:
       i) 1 second   ii) 1 minute?
    b) In alternators in the U.S.A., the electromagnet spins 3600 times each minute. What is the frequency of the a.c. there?
5   **Try to find out:** about the turbines which drive the generators.

Did you know?

● The electromagnet in an alternator does not spin in air. Air resistance would slow down the spinning magnet. Instead, alternators are filled with hydrogen, which has a much lower resistance.

● In some alternators, hydrogen gas is piped through the coils to cool them.

After electrical energy has been generated at the power station, it has to be **transmitted** (moved from one place to another). There's much more to this than just connecting up the power station and each user by a very long cable!

Of course, cables are involved. The electrical energy is transmitted round the country using thick aluminium or copper cables which are hung from pylons or buried underground. The cables are made of aluminium and copper because these metals are good conductors. The cables are thick to give low resistance.

Transformers also have a vital part to play. Their job is to change the voltage of the electricity. This has to be done for two main reasons:

1   Electrical energy is transmitted more efficiently at high voltages.
Whenever a current flows through a wire, some electrical energy is converted to heat energy. (You will learn more about this on page 117.) If the electrical energy is transmitted at high voltage, the currents which flow will be small. Then only a small amount of the energy will be wasted as heat. That's why very high voltages are used when electrical energy has to be transmitted over long distances.

2   Different users require electricity at different voltages.
Electrical energy is supplied to your home at 230 V, but factories may need 11 000 V supply, and electrified railways, 25 000 V.

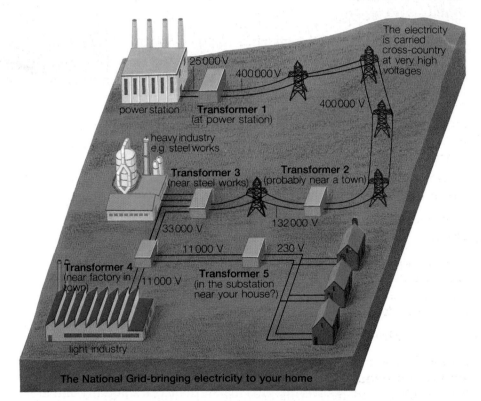

The National Grid-bringing electricity to your home

The network which carries electrical energy round the country is called the **National Grid**. It uses **step-up** transformers to increase the voltage, and **step-down** transformers to decrease it.

1   What is meant by transmitting electrical energy? ▲
2   What are transmission cables made of, and why? ▲
3   What is a transformer's job? What is the difference between a step-up transformer and a step-down one? ▲
4   Which of transformers 1–5 in the diagram are step-up, and which are step-down?
5   Why is electrical energy transmitted at high voltage? ▲
6   a) What is the National Grid? ▲
    b) Where, in the Grid, is the voltage highest, and why?
7   **Try to find out:** where your nearest substation is.

Did you know?

● The Grid system must be able to supply extra electricity at some times of the day (like 5 pm, when everyone makes tea!).
● There are some power stations which are kept working at half capacity so that they can supply this extra electricity.

Electricity drives many moving machines. It drives model trains – and huge passenger ones!

As you might expect, magnetism is involved in this, too. You can show this with a simple experiment. If you put a strip of metal foil between the poles of a magnet, then pass a current through it, the strip moves. If you make the current flow the other way, the strip moves in the opposite direction.

When a current flows through a wire which is in a magnetic field, a force acts on the wire. This force makes the wire move.

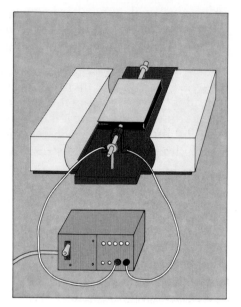

aluminium foil

Electric motors

'Up and down' movement is not really of much use. The turning movement of an electric motor is much more useful.

A simple electric motor is built in the same way as the model generator on page 112. A coil of wire is fixed so that it can spin on an axle between the poles of a magnet. Supplying electricity to the coil makes it turn.

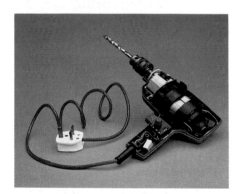

This electric drill has been opened to show the motor inside

The motor of a model electric train

Electric motors are used in many machines like electric drills, food mixers, and washing machines. The motors usually have several coils, each with many turns of wire. This makes them run smoothly and more powerfully. Many motors use electromagnets. These can be much more powerful than permanent magnets of the same size.

1 How can you produce movement using an electricity supply, a wire, and a magnet? ▲
2 What would you need to make a simple electric motor? How would you make it work? ▲
3 What is the energy change in an electric motor?
4 Why do most motors have several coils of wire? ▲
5 List some machines which have electric motors in them.
6 The controller of a toy electric train can change the current fed to the train's motor. Suggest what happens to the current when you make the train: a) speed up b) slow down c) go backwards.
7 **Try to find out:** some disadvantages of electric motors.

Did you know?

- Electric motors have some advantages over fuel burning engines.
- Electric motors are cleaner, and cause no pollution.
- Electric motors are quieter. A milkman driving an electric milk float isn't likely to waken you early in the morning!

15.3 Using electricity – for heating

Whenever an electric current flows through a wire, the wire heats up. Some of the electrical energy is changed to heat energy. The larger the current, the greater is the amount of heat.

This heat can cause problems. Hot wires can cause fires! If the current flowing in a circuit is too large, the heat produced may melt the plastic insulation on the cables or set it on fire.

But hot wires have lots of uses. The rest of this page deals with appliances which **safely** change electrical energy into heat.

Inside each ring of an **electric cooker** is a heating wire. It changes much of the electrical energy to heat energy.

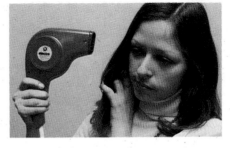

There are coils of heating wire inside this **hair drier**. A fan blows air over the wires, and the hot air dries your hair.

Inside an **electric toaster**, there are lots of heating wires. You can see them if you look inside, but make sure that you unplug the toaster first.

This is an **oil-filled radiator**. Heating wires run inside it. The wires heat the oil, then the oil heats the surrounding air.

Running through an **electric blanket** there are many thin heating wires.

Some rugby and football pitches have their own **'electric blankets'**. Heating wires run through the soil near the surface.

1 What energy change takes place when a current flows in a wire? ▲
2 Why does a television get hot after being on for some time?
3 Which of the heaters (shown above) heat their surroundings by:
   a) conduction b) convection c) radiation?
4 If a current of 30 A flows through a circuit which is designed to take 15 A, there is a danger. What is it?
5 Suggest why the electric drill (on page 116) has a fan for blowing air over the motor.
6 Why should a rugby pitch need an electric blanket?
7 **Try to find out:** some safety rules for using an electric blanket.

Did you know?

- Electromagnets which use high currents can get very hot. Some have special cooling systems to get rid of the heat.
- About 60 km of heating wire run through the top soil of Scotland's rugby pitch at Murrayfield, Edinburgh.

15.3 How much energy has to be paid for?

Going further

Have you ever been in trouble for leaving lights on, or for having the TV on too long, or for forgetting to switch off an electric fire? Of course you have! Everyone has!

You really should make a point of switching off an appliance when you have finished using it. Using electrical energy costs money! But some appliances are more expensive to run than others. That's because they need more energy to keep them working.

The amount of electrical energy used depends on two things:

1 The time for which the appliance is working. The longer an electric fire (or TV, or light bulb) is on, the more energy it uses. It uses twice as much energy in 2 hours as it does in 1 hour.

2 The power of the appliance. From the power of an appliance, you can tell how much energy it uses in 1 second. Power is measured in **watts** (W) and **kilowatts** (kW). In 1 second, a 500 W toaster uses 5 times more energy than a 100 W light bulb. A 1 kW fire uses even more energy – 10 times more than the bulb. (1 kilowatt = 1000 watts.)

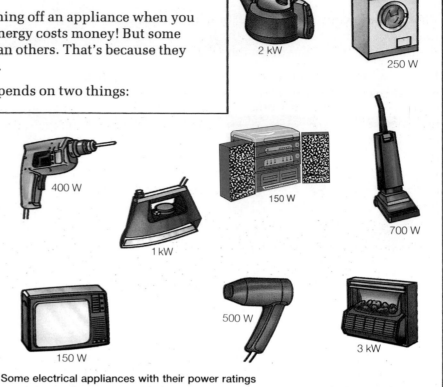

Some electrical appliances with their power ratings

Appliances normally have their power rating marked on them. If you know the power rating, and the length of time for which the appliance has to be working, you can find the energy used from:

> **electrical energy used = power × time**

The Electricity Board charges for **units** of electrical energy called **kilowatt hours**.

Number of units of electrical energy used = power × time
(in kilowatt hours)   (in kilowatts)   (in hours)

If a 2 kW heater is switched on for 4 hours:

number of units used 2 × 4 = 8

1 What can you tell from the power of an appliance? ▲
2 a) Complete the bar chart for the appliances shown above.
  b) Which use more electrical energy, appliances for heating or moving?
3 The amount of electrical energy used by an appliance depends on two things. What are they? ▲
4 How can you find the number of units used by an appliance? ▲
5 Work out the power (in kW) of: a) the TV set b) the light bulb.
6 How many units are needed to run: a) a 3 kW bar fire for 6 hours b) a TV set for 6 hours c) a light bulb for 24 hours?
7 Which is more expensive – leaving on a light, a TV or a bar fire?

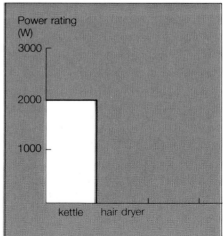

Power rating (W)

Did you know?

● Cookers are rated at around 6–8 kW. That's why a cooker has to be wired into a circuit of its own, with a large fuse.

15.3　What size of fuse?

When you buy a new television, or stereo, or hairdrier, you usually have to buy a plug as well. Most plugs are bought ready fitted with a 13 amp (13 A) fuse. Unfortunately, a 13 A fuse is quite unsuitable for many household appliances (including a television, stereo, and hairdrier!). Your first job will probably be to change the fuse.

What does a fuse do?

Most of the time a fuse does nothing, except allow a current to flow through the plug to the appliance. The current is carried across the fuse by a piece of fuse wire. The fuse wire in a 13 A fuse allows a maximum current of 13 A to flow, the fuse wire in a 5 A fuse allows a maximum current of 5 A to flow, and so on. If a larger current does flow, the fuse wire melts, breaking the circuit and stopping the flow of current. Then the fuse has done its job, which is to prevent too large a current from flowing.

What size of fuse should be used?

Shops normally stock 1 A, 3 A, 5 A and 13 A fuses. The fuse you fit in an appliance's plug should:

- be able to carry the current which the appliance normally uses
- melt if too large a current flows.

If a dryer uses a current of 2 A, there's no point fitting a 1 A fuse in its plug. The fuse wire will melt whenever the dryer is switched on! Fitting a 13 A fuse is just as silly. If a fault develops and a large current starts to flow, the fuse won't melt until the current reaches 13 A. That size of current can do lots of damage. A 3 A fuse is correct.

How can you tell the size of the current used?

Unfortunately, the current used by an appliance is not normally marked on the appliance. It is, however, possible to work out the current from the appliance's power.

Current used (in amps) = $\dfrac{\textbf{power} \text{ (in watts)}}{\textbf{voltage} \text{ (in volts)}}$

In Britain, the voltage is 230 V, and so the current used by a 2 kW (2000 W) electrical kettle is given by:

Current = $\dfrac{2000}{230}$ watts = 8.7 amps　　Correct fuse is 13 A

1　What is a fuse's job? How does it do this job? ▲
2　The fuse you fit in an appliance's plug should do what? ▲
3　A TV uses 1.5 amps. Which fuse should be used and why?
4　How can you work out the current used by an appliance? ▲
5　Which of the appliances on page 118 uses the largest current. Explain your choice, but don't do any calculations.
6　From the power ratings given on page 118, work out the current used by:　a) the iron　b) the bar fire　c) the TV. Choose the correct fuse for each one.

Which of these fuses....

should be put in here?

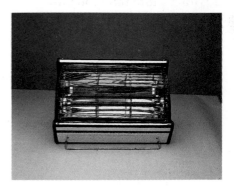

That depends on the current used by the appliance. You can work that out if you find this

119

Index

Acknowledgements

The publishers would like to thank the following for permission to reproduce photographs:

B & C Alexander: p.88 to right; Allsport: pp.76 centre right, 78 top, 79 bottom, 83 top, 86 centre; Aspect Picture Library: pp.52 top, 83 bottom; Associated Press: p.10; Barnaby's Picture Library: pp.40 centre, 48 top, 56 centre, 93 lower left, 98 top, 110 left; Anne Bolt: p.61 inset; Paul Brierley: pp.46 bottom left and centre, 53 top right, 56 top and bottom right, 57 centre left, 66 centre; British Aerospace: 21 bottom; British Geological Survey: pp.46 top right, 56 left, 57 top right; British Gliding Association: p.6 top; Camera Talks: pp.88 top left, 91 top centre, 93 lower centre; Cement and Concrete Association: p.53 centre right; Central Electricity Generating Board: pp.114 right, 115 top; Bruce Coleman/Jane Burton: pp.31 left, centre and right, 38 bottom centre and right; Bruce Coleman/ B and C Calhoun: p.38 top left; Bruce Coleman/Thompson/Lane: p.38 bottom left; Bruce Coleman/A. Compost: p.67 bottom; Bruce Coleman/H. Reinhard; p.94 top inset; Bruce Coleman/Nocholas de Vere: p.55; Bruce Coleman/J. van Wormer: p.105 top; Bruce Coleman/M. P. L. Fogden: p.105 centre; Brice Coleman/M. Kahl: p.105 bottom; Controller of HMSO (Meteorological Office):p.16 top; Crown Copyright Central Office of Information: p.110 right; Telegraph Colour Library/Ian Vaughan: p.18; Electricity Council: p.117 top left, top centre, bottom centre; Mary Evans Picture Library: p.15 left; Forestry Commission: p.34 bottom; Alex Fraser/OUP ©: pp.4 top right, bottom left, 19 top, 20 top, centre right, bottom left, 23 centre, 28, 40 left and right, 46 top left and centre, 50 top left, 51 top left, 63 top right, 66 centre right, 67 centre, 68, 70, 72, 76 bottom left, 77 top, 84 bottom, 96 bottom, 109 left, 111 top, 117 bottom left, 119 middle; Geoscience Features: pp.46 bottom right, 47 top, 49 top, 51 top right and bottom left, 57 lower left, 65 top right; G.S.F./University Film Service: p.7 right; Sally and Richard Greenhill: pp.64 top left, 88 bottom right; Peter Gould/OUP ©: p.27 bottom left and right; Chris Honeywell/OUP ©: pp.3 bottom left, 9 inset, 21 top, 23 left and right, 37 centre, 64 centre and right, 76 centre right, 86 left, 88 bottom left, 98 left, 108 top, 109 right, 116 bottom left and right, 117 top right, 119 top, second from top and bottom; John Radcliffe Hospital, Oxford: pp.91 top right, 94 bottom; Mary Rose Trust: p.20 centre left; Mercedes Benz: p.18 top; Mountain Camera/John Cleare: pp.3 bottom centre, 12 centre; Mountain Camera/Cameron McNeish: p.76 bottom right; NASA: pp.12 bottom, 15 right, 45 top; Oxford Eye Hospital, Radcliffe Infirmary: p.110 centre; Oxford Scientific Films/Bob Fredrick: p.38 top right; N.E.I. Parsons: p.114 left; Picturepoint Ltd: pp.7 left, 49 bottom, 52 centre and centre right, 54 top and bottom, 93 lower right, 104 left and centre; Pilkington Brothers plc: pp.1, 4 centre; Potterton Commercial/Bailey Lauzel: p.14; Stuart Robertson: p.12 top; St Bartholomew's Hospital, London: pp.20 bottom right, 102 bottom; Science Photo Library: pp.9 top and key, 76 top, 98 centre and right, 99 top; Sealed Air Ltd: p.3 bottom right; Sheridan Photo Library: 91 top left, Adrian Smith: p.113 top left; Sportapics Ltd: p.117 bottom right; Tony Stone Images/David Woodfall: p.58; Tony Stone Images/Ernst Ziesmann: p.27 top left; Supersport Photographs/Eileen Langley: p.86 right; Thames Valley Police: p.38 top centre; Topham Picture Library: pp.3 top right, 94 middle inset; United Kingdom Atomic Energy Authority: p.62 top; J.D. Walton Jnr.: p.13 top; ZEFA Picture Library: pp.17, 27 top left, 50 top right, 51 bottom right, 52 centre left, 66 top right and centre left, 67 top, 91 bottom.

With special thanks to Thermos Ltd. and Howes of Oxford.

Illustrations by Bob Chapman, Kate Charlesworth, Roger Gorringe, Paul Thomas/Techniques.

: . . . . and thanks

We should like to thank

- the readers for considered and constructive criticism of the text
- friends and colleagues for time spent discussing and trying out the material in this book, and for helpful comments and suggestions and, most especially,
- wives and families for long suffering, understanding, support and encouragement, without which this book would not have progressed beyond the first few pages.